Growing and Showing Roses

Don Charlton

David & Charles
Newton Abbot London North Pomfret (Vt)

Sincere thanks to Mrs Lillian Benson for typing the manuscript, Maurice Benson for taking the black and white photographs, and to my long suffering family for their help and constant encouragement.

British Library Cataloguing in Publication Data

Charlton, Don
Growing and showing roses. –
(Growing and showing)
1. Rose cultivation
I. Title II. Series
635.9'33372 SB411

ISBN 0–7153–8576–3

Photoset in Souvenir by
Northern Phototypesetting Co, Bolton
and printed in Great Britain by
Redwood Burn Ltd, Trowbridge, Wilts
for David & Charles (Publishers) Limited
Brunel House Newton Abbot Devon

Published in the United States of America
by David & Charles Inc
North Pomfret Vermont 05053 USA

Contents

Introduction

Growing and exhibiting roses is a friendly, interesting and challenging hobby. The opportunity to participate is open to all rose growers, regardless of the number of bushes they have.

The following chapters provide a step-by-step guide to the selection, cultivation, protection, cutting, transportation and staging of exhibition roses. The information should enable you to win prizes – whatever the competitive level – if you wish to work hard with your roses.

Personal experience and analysis has led me to believe that there are deficiencies in the accepted methods of growing show-standard roses and that we need to introduce new ones. My own tried and tested improvements have been included here – and I hope they will enhance the winning potential of existing exhibitors, and attract and encourage newcomers to the hobby.

Rose exhibitors usually start at small local shows, and progress via larger horticultural shows or specialist affiliated rose society shows. Confidence gained at these levels often leads to involvement later in national competition, where shows are organised by the Royal National Rose Society. I strongly recommend that you join your local rose society, for then a wealth of experience and knowledge will be available to you. Joining the Royal National Rose Society will also give you access to a number of publications covering many aspects of growing and showing roses – write to The Secretary, The Royal National Rose Society, Chiswell Green Lane, St Albans.

Having decided to compete, the next step should be to consider any limiting factors, eg the maximum number of bushes you can grow, and the availability of time and transport. Then you should decide upon your aims. Providing you always work within your limits, and use the relevant information in this book, you will be following a path that should lead to success! However, as in all other competitive activities, results depend not only upon your skill, ability and knowledge, but also upon your level of commitment, which often proves decisive.

Never be afraid to compete – if you do not try, then you can never succeed. At a large local show where I exhibited, there was always staging set up with empty vases placed upon it. Nearby stood a notice which read 'Reserved for all those people who may have better blooms in their gardens at home'!

At this point I should stress that the aims and methods gradually outlined and developed in this book are my personal opinion and, of course, open to interpretation, improvement and development. These methods have, however, proved extremely successful and may provide a starting point for others.

1 Classification and Varieties

Before deciding which types of rose to grow and exhibit, it is important to understand rose classification. The new system approved and introduced in recent years by the World Federation of Rose Societies has hopefully brought some order – previously, individual countries were instituting their own classifications and groupings. The new class names have come into use gradually in rose literature and show schedules, and are usually followed by the old descriptions in brackets. Many older books, together with some present catalogues and schedules, continue to use the old descriptions. Roses are here divided into two groups for show purposes: Modern Garden roses and Old Garden roses.

Modern Garden Roses

These include Climbing, Shrub, Bush and Miniature roses. The Bush roses are sub-divided as follows:
Large Flowered (Hybrid Tea type or HT) roses – Specimen Blooms
Large Flowered (Hybrid Tea type or HT) roses
Cluster Flowered (Floribunda type) and Polyantha (Pompon type) roses

Old Garden Roses

These include roses already established in classification before the introduction of Hybrid Tea roses in the 1860s (a result of crossing Hybrid Perpetual roses with Tea roses).

Selecting Varieties for Exhibition

Almost all rose varieties can be used for exhibition, but only a small proportion of them can be relied upon regularly to produce blooms of the required quality and characteristics, which will stay in peak condition for a reasonable time. Selecting such varieties is therefore a very important factor for exhibitors. The qualities required for a good exhibition variety (plus the faults to avoid) are given with the following detailed descriptions of each type.

Large Flowered (HT) Roses – Specimen Blooms

These must be large examples of classic blooms (of any particular variety), at the peak of development. The bloom should be at least three-quarters open, circular in outline, displaying a regular, conical, high-pointed centre, fresh in appearance, with bright colouring typical of the variety, and possessing petals of good texture (not too coarse or thin and papery). The bloom should be unblemished by weather or insects (see page 8).

Large Flowered (HT) Roses

These are normal-sized blooms of a particular variety. They should be half to three-quarters open with characteristics similar to those described for specimen blooms. Usually, these younger blooms display enhanced freshness and colour.

These and specimen blooms may be cut from the same bush. Size and development are the only differentiating factors. Climbing and shrub forms of Large Flowered roses may be exhibited under these headings.

Exhibition qualities and faults
Varieties should possess the ability to produce a reasonable number of blooms at each flush, and the blooms should stay in peak condition for a reasonable time (12–24 hours). Blooms should not open too quickly when being transported or lose colour quickly after cutting. Other qualities required are: healthy foliage, well-balanced stems and an ability to take up water.

Many desirable varieties regularly show the two most common bloom faults: split and snub-nosed centres. Either fault rules them out for exhibition purposes. A bloom with a split centre has a marked division in the centre of the bloom, instead of the desired high-pointed, conical centre. In bad cases this can appear as two centres joined by a few petals. Snub-nosed centres have centres with flat tops (see opposite).

Cluster Flowered (Floribunda type) Roses and Polyantha (Pompon type) Roses

These include roses described previously as floribunda, floribunda HT type, floribunda shrub, cluster flowered climbers and ramblers. Flower heads should be large, circular and reasonably flat, with individual blooms adequately spaced to allow full development without overcrowding. Blooms should have substance and freshness, together with brilliance and purity of colour.

Exhibition qualities and faults

The requisite qualities are an ability to produce consistently stems carrying large, weather-resistant bloom heads. The blooms' colour should not fade in sunshine. Stems should be of reasonable length (38–66cm or 15–26in) and capable of taking up water. This latter quality is important, as stems often carry numerous blooms and considerable foliage. Foliage should be fresh and healthy. The most common faults are lack of weather resistance and blooms which fade quickly.

Miniature Roses

Bush and climbing Miniature roses are characterised by miniature growth of stems, foliage and blooms. Blooms may be shown singly or in clusters. They range in form from single to very double (possessing from five to over forty petals). Standards of substance, freshness, brilliance and purity of colour are as for the full-sized roses.

Bloom with split centre

Bloom with snub-nosed centre

Weather-damaged bloom

Perfect bloom

Exhibition qualities and faults
Blooms (single or clusters) should be well shaped, adequately spaced and fairly weatherproof. Colours should hold for a reasonable time and foliage should be healthy. Common faults include blooms which open or fade too quickly.

Old Garden Roses

These are the Alba, Bourbon, Boursault, China, Damask, Gallica, Hybrid Perpetual, Moss, Portland, Provence and Sweet Briar roses. They produce stems, foliage and blooms displaying a multitude of characteristics. All are suitable for exhibition. Hips and thorns also improve the display.

Exhibition qualities and faults
Blooms may vary greatly in size, shape and form, and all are acceptable. Good stems and healthy foliage are desirable, though some older varieties are martyrs to disease and blooms can be brittle or have weak necks.

Recommended Exhibition Varieties

The next step is to look at the varieties which are currently successful and frequently used by winning exhibitors. Careful choice from the following lists will produce a reliable basis for your exhibition collections. Several of the Large Flowered varieties (termed 'bankers') are regarded as indispensable and are listed separately. Almost all Cluster Flowered and Miniature roses can be exhibited, though I have recommended some for guidance.

No specific recommendations are made about Old Garden roses – many suitable varieties are listed in specialist catalogues.

Recommended Large Flowered varieties (bankers)

Variety	*Colour*	*Comments*
ADMIRAL RODNEY	Pale mauve, deeper petal markings	Large blooms, hold well. Often has curved stems. Dislikes rain.
BOBBY CHARLTON	Deep pink, paler reverse	Blooms excellent for all purposes. Fairly weatherproof. Tends to bloom late.
CITY OF BATH	Pink, paler reverse	Smaller blooms, hold well. Weather-resistant.
CITY OF GLOUCESTER	Pale yellow touched with pink	Shapely blooms, sometimes frilled. Hold well. Dislikes rain.
FRED GIBSON	Pale to deep cream/gold	Blooms excellent for all purposes. Some weather-resistance.
GARY PLAYER	Cerise-pink and white	Blooms may be frilled, hold well. Almost weatherproof.
GAVOTTE	Pink, paler reverse	Blooms shapely, hold well. Dislikes rain. Stems often curved.
GRANDPA DICKSON	Pale lemon, yellow	Blooms shapely. Weather-resistant. Colour varies by location. Best yellow.
RED DEVIL	Deep scarlet, turkey reverse	Blooms shapely. Good for all purposes. Stems and foliage excellent.
ROYAL HIGHNESS	Blush-pink	Blooms shapely, reliable, good for all purposes. Dislikes rain.
RED LION	Deep pink	Blooms shapely. Weather-resistant.

Other recommended Large Flowered varieties (by colour)

Variety	*Colour*	*Comments*
MEMORIAM	Blush-pink	Useful when twelve or twenty-four different varieties are required for box exhibits. Dislikes rain.
SILVER LINING	Pink, paler reverse	Beautiful bloom. Useful where twelve or twenty-four varieties required.
ANNE LETTS	Pink, paler reverse	Shapely variety but prone to split centres. Dislikes rain.
ISOBEL ORTIZ	Pink, paler reverse	Some huge blooms. Only grow for box classes. Weather-resistant.
SILVER JUBILEE	Pink and cream	Shapely blooms. Weather-resistant. Good reliable variety.
DREAMTIME	Pale pink	Well-shaped blooms, last well. Tends to flower early.
BONSOIR	Apple-blossom, pink	Very reliable in some areas. Lovely colour. Dislikes rain.
PETER FRANKENFELD	Deep pink	Borders on being a banker. Good shape. Weather-resistant.
LILLY-DE-GERLACHE	Deep pink, copper veining	Large blooms on long stems. High proportion of confused centres.
PEACEFUL	Deep pink	Old variety suitable for boxes. Weather-resistant.
KATHLEEN O'ROURKE	Orange-pink	Produces shapely blooms.
FIRST PRIZE	Pink and cream	Huge shapely blooms. Dislikes rain. Good box variety.
OXFAM	Cerise-pink	Blooms good shape. Hold well. Some split centres.

SHOWGIRL	Cerise-pink	Very old variety. Good proportion of shapely blooms.
KEEPSAKE	Cerise-pink	New variety. Could be a future banker. Very shapely. Hold well.
MY JOY	Cerise-pink	Sport from Red Devil, similar characteristics. Possible banker.
NORMAN HARTNELL	Carmine-pink	Large blooms. High proportion of split centres. Weatherproof.
HARVESTER	Pale mauve, deeper markings	Useful for multi-variety box classes. Shapely. Some split centres.
STEPHANIE DIANE	Pale red	Good blooms. Resists rain.
BIG CHIEF	Deep scarlet	Beautifully shaped blooms. Some split centres. Resists rain.
GOLIATH	Plum-red	Useful for multi-variety box classes. High proportion of split centres.
HOT PEWTER	Vermilion	Beautiful colour. Weather-resistant blooms.
CHAMPION	Yellow and scarlet	Beautiful, weather-resistant blooms. Prone to splitting.
EMBASSY	Pale yellow/gold, red veining	Beautiful large blooms. Suits cooler climates.
JAN GUEST	Gold and scarlet	Produces well-shaped blooms. Some weather-resistance.
GARDEN PARTY	Pale yellow, pink markings	Blooms sometimes frilled. Hold well. Useful for multi-variety box classes.
BOB WOOLLEY	Pale yellow, pink shading	Large blooms but prone to splitting.
DOVEDALE	Cream and red	Beautiful colour. Fairly weather-resistant. Bred by an amateur.

Recommended Cluster Flowered varieties

Variety	*Colour*	*Comments*
ANNE HARKNESS	Saffron, apricot	Large well-shaped bloom heads on good stems. Weather-resistant.
ANNE COCKER	Orange, scarlet	Bloom heads variable in shape. Beautiful colour. Great weather-resistance.
CITY OF LEEDS	Salmon-pink	Well-shaped bloom heads. Holds colour. Good weather-resistance.
DOROTHY WHEATCROFT	Orient-red	Very reliable. Great weather-resistance. May need some bud thinning.
DREAM WALTZ	Blood-red	Good bloom heads. Velvet petal texture. Weather-resistant.
EUROPEANA	Deep blood-red	Huge heads of blooms. Weatherproof. Stems may be curved. Prone to mildew.
FRED LOADS	Vermilion, orange	Good flower shape, profuse bloomer. Vigorous grower. Stands some rain.
GRACE ABOUNDING	Cream-white	Good bloom heads. Lovely colour. Dislikes rain.
ICEBERG	White	Good head shape. Profuse bloomer. Dislikes rain.
LIVERPOOL ECHO	Soft salmon-pink	Good bloom shape. Large heads, good stems. Dislikes rain.
MATANGI	Vermilion, white eye	Good bloom heads. Startling colour. Weatherproof.
SALLY HOLMES	Creamy-white	Huge bloom heads, beautiful colour.

Recommended Miniature varieties

Variety	*Colour*	*Comments*
STARINA	Vermilion-scarlet	Excellent. Good colour. Weather-resistant. HT-shaped blooms.
SHERI-ANNE	Orange-red	Good variety. Weather-resistant. HT-shaped blooms.
FIRE PRINCESS	Deep orange-red	Lovely HT-shaped flowers. Very vigorous.
RED ACE	Crimson	Very good variety. Shapely trusses of flowers, splendid colouring.
DARLING FLAME	Orange/red, yellow	Good variety. Keeps its colour.
RISE AND SHINE	Medium yellow	Good colour. HT form. Hold well.
JUDY FISCHER	Deep pink	Excellent variety. Profuse bloomer. Holds colour well.
STACEY SUE	Mid-pink	Good variety. Beautiful colour. Holds form well.
ELEANOR	Rose-pink	Good variety. Holds colour.
SNOWDROP	White	Good flower trusses. Beautiful when flowers wide open.

Number of Bushes

After examining the types and varieties, you must decide whether to compete generally or specialise in one particular area, and how many bushes to grow. Start with a number which you can handle comfortably and build up from there. There are no restrictions on the number of bushes you can grow, but schedule divisions are usually made on the basis of the number of bushes, to ensure fair competition. Always consult the show schedule's requirements.

Let us assume that 250 bushes are to be grown – divided between Large Flowered and Cluster Flowered types. A typical requirement to cover all classes would be a total of 40 stems of

Large Flowered roses in at least 4 different varieties, plus 10 stems of Cluster Flowered in at least 2 varieties. Where 500 bushes are grown, the schedule demands 60 stems of Large Flowered roses in 8 varieties, and 20 stems of Cluster Flowered roses in 4 varieties. This illustrates how the degree of difficulty increases with a greater number of bushes.

The original total of 250 bushes should be broken down into 210 Large Flowered and 40 Cluster Flowered bushes. Such a ratio will usually provide the number of blooms or trusses required to cover schedule demands. For the Large Flowered varieties, first select six 'bankers' and add other varieties from the recommended list. Experienced exhibitors grow their chosen varieties in sufficient numbers to provide for show requirements over the whole season.

Before finalising your choice, try to visit local exhibitors and rose nurseries to check on the health and vigour of your chosen varieties. Then visit shows to see these varieties exhibited in boxes, vases and bowls. Remember, the final object is to produce sufficient blooms in the required number of varieties to enable you to at least compete in all classes.

I have made selections of Large Flowered and Cluster Flowered varieties, giving two possible breakdowns for each type. Choice A or B will give the required number of bushes and a good range of colours, covering the whole season. Having a second breakdown gives scope for a different colour range.

Miniature roses have become extremely popular over the last five years. Some exhibitors specialise in them; they take up much less room, provide a colourful display and their blooms are fairly weatherproof. Miniature bushes are not included in normal schedule-division total calculations – they are assessed separately.

There are no limitations on the number of Old Garden roses grown.

Selection of Large Flowered varieties (210 bushes)

Variety	*Colour*	*No of Bushes (A)*	*No of Bushes (B)*
RED DEVIL	Deep scarlet/turkey	30	20
ROYAL HIGHNESS	Blush-pink	30	20
GAVOTTE	Pink, lighter reverse	30	20

FRED GIBSON	Pale cream/gold	30	20
GRANDPA DICKSON	Pale lemon-yellow	20	20
ADMIRAL RODNEY	Pale mauve, deeper markings	20	20
CITY OF BATH	Pink, lighter reverse	10	10
BOBBY CHARLTON	Deep pink, lighter reverse	10	10
GARY PLAYER	Cerise/white	10	10
CITY OF GLOUCESTER	Pale yellow/pink	0	10
SILVER JUBILEE	Pink and cream	5	6
JAN GUEST	Gold and scarlet	5	6
BIG CHIEF	Deep scarlet	5	10
KEEPSAKE	Cerise-pink	5	8
PETER FRANKENFELD	Deep pink	0	8
EMBASSY	Pale yellow/gold	0	6
HOT PEWTER	Vermilion	0	6
		210	210

Selection of Cluster Flowered varieties (40 bushes)

Variety	*Colour*	*No of Bushes (A)*	*No of Bushes (B)*
DOROTHY WHEATCROFT	Orient-red	8	6
FRED LOADS	Vermilion-orange	8	6
ICEBERG	White	8	6
ANNE HARKNESS	Saffron-apricot	8	6
EUROPEANA	Deep blood-red	8	6
LIVERPOOL ECHO	Soft salmon-pink	0	5
GRACE ABOUNDING	Creamy-white	0	5
		40	40

2 Stocks and Propagation

Order your bushes as early as possible, stating your requirements and delivery date clearly. Ordering during July, August or September usually guarantees delivery without disappointment.

The basic sources for bare root roses are specialist nurserymen, local nurserymen, garden centres and supermarkets. Supermarkets should be avoided because the roses are usually packaged, are not always top grade and are stored in conditions which often force them into premature growth. The other three sources are all acceptable, but local nurserymen or garden centres can rarely provide a comprehensive list of exhibition roses.

The specialist nurseryman is the main source of supply, but, even so, you may have to split your orders between several suppliers in order to obtain all the required varieties. These nurserymen have built up good reputations over the years, and their livelihood depends on their top-class product. Some recommended suppliers are: Dickson of Newtownards, Northern Ireland; Sandays of Bristol; Fryers of Knutsford; Warley Roses of Warley; Gregory's of Nottingham; Gandy's of Lutterworth; and a specialist exhibition rose supplier F. Haynes, based in Kettering.

Home Propagation

There are certain advantages in producing your own bushes by budding on to understocks. Understocks are very much cheaper than bushes. They can be planted in their final growing positions and budded *in situ*, thus avoiding the transplanting stage. Desirable (but costly) new varieties, which are in short supply, may be propagated at a negligible cost, once the initial bushes have been obtained from nurserymen. Finally, growers can experiment with any one of the understocks available and establish the best type for local soil and conditions.

There are also disadvantages. There is a delay of about eighteen months between planting stocks and obtaining blooms, with only a limited number of blooms being produced in the first season. During this time, these bushes occupy land which could be producing blooms. Budding rose understocks successfully depends on weather conditions and the skill of the budder; amateurs may have a low success rate if the weather is dry. Also, there can be a shortage of budwood. Finally, budding takes place during the rosegrower's busiest period.

If you have time on your hands, spare growing space and a limited budget, then home propagation is an excellent idea. Indeed, most exhibitors try budding their own understocks at some time. Rose understocks (usually Laxa, although other varieties are available) are obtained from specialist nurserymen and planted about 20cm (8in) apart, from January to March. The necks should be just covered by soil.

Rootstocks are mainly budded during the following July and August (though this can continue into early September), after periods of rain, or two to three days after the understocks have been thoroughly watered. The moisture encourages the sap to flow and increases the chance of the buds taking.

The budding operation

Obtain 30cm (12in) stems of varieties to be budded on; they should have carried perfect flowers if possible. Cut back all the leaves, leaving only short leafstalks to act as handles later, and remove all thorns. The thorns should snap off cleanly if the stem is ripe enough to be used for budding. From the centre of each stem, select two or three buds which are not too advanced or retarded. They should be obvious, but should not have started to form new stems.

Remove the buds using a knife with a thin, very sharp blade. Insert the blade under the stem bark, about 1cm (½in) above the selected bud, guide down and under the bud, cutting just into the stem wood, then bring the blade to the surface of the bark, about 2cm (¾in) below the bud. If the bark is not cut through, tear downwards until it is freed.

This procedure usually leaves some stem wood attached behind the shield of bark containing the bud. Remove this using both hands; one gripping the bud shield which has the leafstalk attached, and the other holding the top of the stem wood (after peeling the bark away). Hold the shield firmly but gently, and

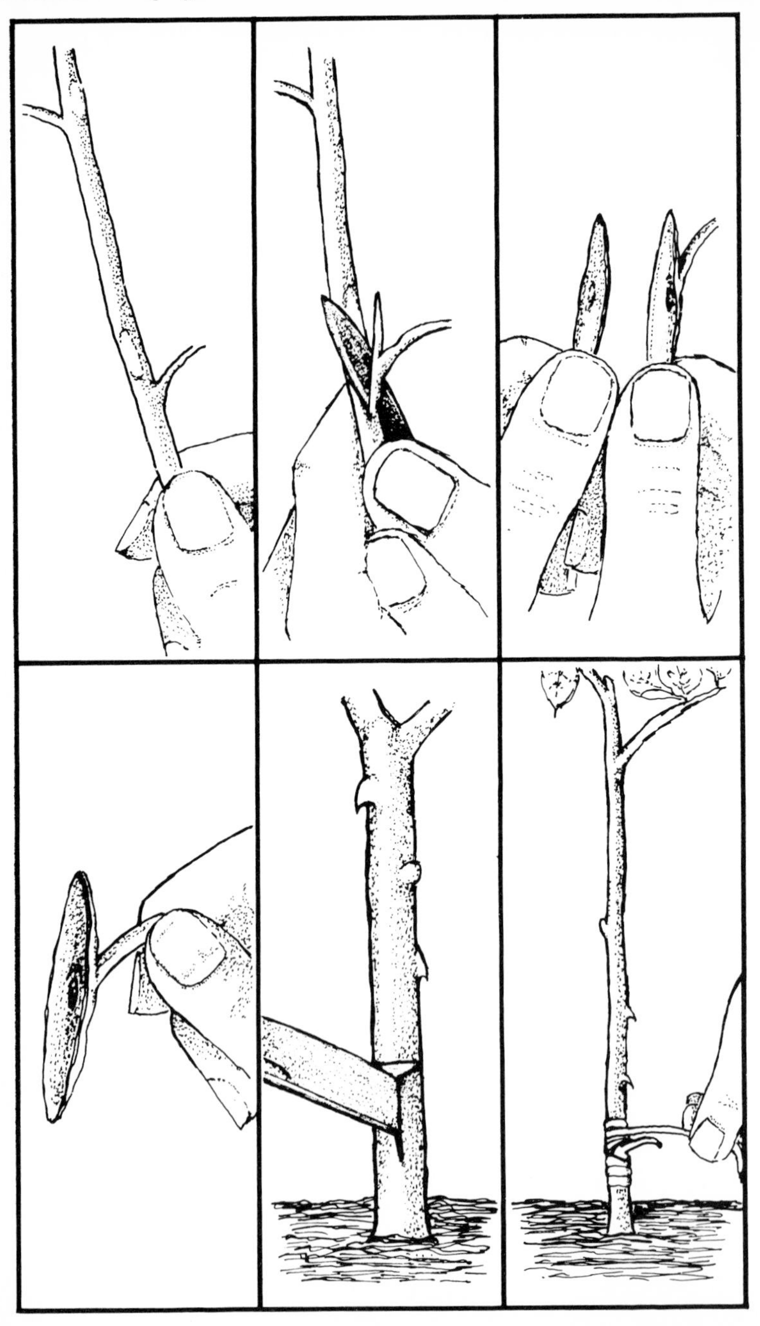

The budding operation

snatch the stem wood quickly downward with a twisting motion. Check that the dormant bud remains attached to the bark shield. (Practise this operation well before budding day.) Keep the budding shields with their eyes in water.

Turning to the understocks, soil must now be cleared from around their necks, where the buds will be inserted. Clean the necks thoroughly with a damp cloth; this usually reveals clean white bark. Make a 'T' cut just through the bark, the vertical incision being 2.5cm (1in) long and the horizontal incision 1.5cm ($\frac{5}{8}$in). Trim the bud shield to a suitable size. Gently, lift the two flaps of bark on either side of the vertical incision and insert the bud shield. Next, trim any surplus bark from the top of the shield by running the knife blade along the horizontal incision. Secure the bud in position using patent-rubber budding ties, or damp raffia tied around the stem – or I have used thin strips of elastic band. After four weeks, check that the bud is still green and not withered. If it has dried out, bud the understock on the opposite side.

Replace the soil around the understocks until the inserted buds are just at soil level. The following February, understocks have their tops cut off completely, just above the inserted buds: this is called heading back. Buds should then grow away and flowers should form on new wood in July.

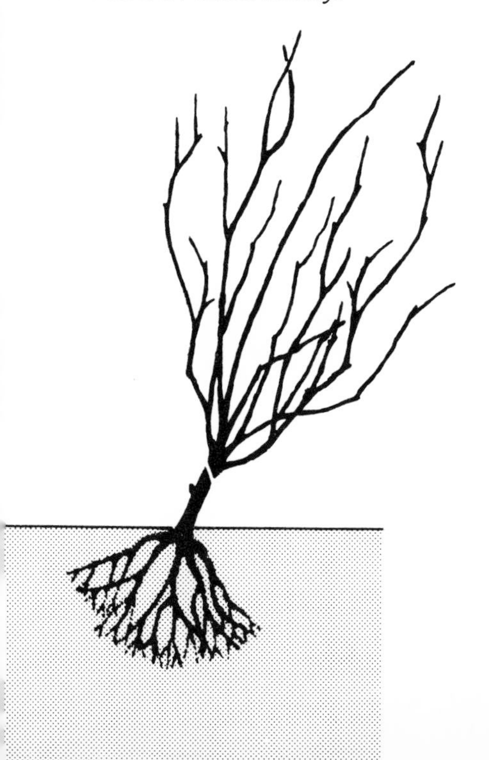

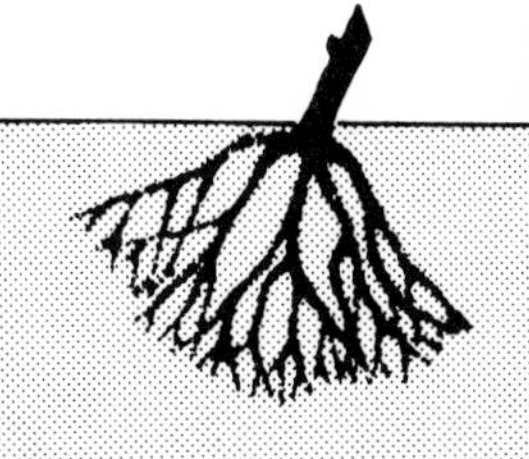

Heading back understocks

3 Planting and Winter Treatment

To produce top-class exhibition roses regularly you must follow a cultivation routine conscientiously throughout the year, beginning with good soil preparation before planting.

Roses prefer to grow in a slightly acidic soil, pH 5.6–6.8, though they will grow in soils between pH5 and pH8. When the soil pH is too low (too acidic) add chalk in the form of garden lime, and when it is too high (too alkaline) add flowers of sulphur or peat. If you are in doubt about pH and soil fertility, arrange for a soil sample test to be made through a specialised laboratory (or use a kit for speed). Analysis is usually available via county agricultural or horticultural organisations, at a reasonable cost.

Humus should be provided by adding garden compost, well-rotted farmyard manure, straw, peat or waste hops. Nourishment should be provided by adding organic fertilisers (eg fishmeal, bonemeal, hoof and horn) and/or inorganics, in the form of proprietary rose fertilisers, which are usually balanced to provide all the elements essential for growth. The three main requirements are nitrogen to sustain good healthy leaves and stems; phosphorus to build up a healthy, vigorous root system; and potassium to ripen stems and provide good bloom colour. Roses also require magnesium, manganese, iron and calcium in smaller quantities.

My own favourite inorganic fertiliser is Enmag which is produced by Scottish Agricultural Industries and combines nitrogen, phosphorus and potassium with added magnesium. Enmag is a special slow-release fertiliser, which releases its feeding elements into the soil over a prolonged period. A dressing applied during the spring will provide continuous nourishment throughout the growing season – saving time and effort. Also, unlike some highly soluble fertilisers, this type is not easily washed through the soil and past the roots during heavy rain.

Powdered seaweed (Seagold) can enhance bloom colour, and helps to build up a good soil crumb structure.

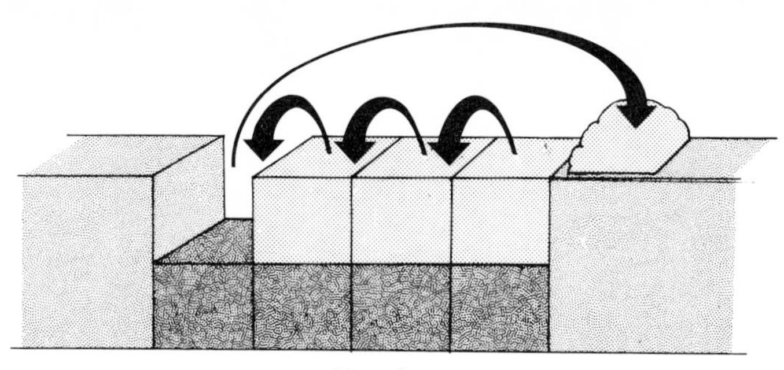

Trenching

Before a plot of ground is used for rose growing, the whole area should be cultivated, not just the actual beds. This discourages water from draining into the bed from heavy surrounding soil. It also encourages rose roots to spread under pathways.

The best method of cultivation is bastard trenching. Mark out the whole area of land in strips 46cm (18in) wide, running across the plot. Dig out the first strip to a depth of 23cm (9in), then use a wheelbarrow to move the soil away and tip it at the opposite end of the plot.

Next, the soil at the base of the first trench should be dug over and broken up. If the soil is a heavy clay, add some gypsum (calcium sulphate) at the rate of 0.8–1.0kg per metre (1½–2lb per yd) of trench. This produces larger soil crumbs, which aids drainage and allows better aeration.

Add well-rotted farmyard manure, garden compost, or rotten straw, at the rate of a large barrow-load per 1.8m (2yd) of trench. Other nourishment such as steamed bonemeal, fishmeal, powdered seaweed and a proprietary rose fertiliser (preferably slow-release) should be added at a rate of 63–94g per metre (2–3oz per yd) of trench. Mix all additives well into the soil using a spade or fork.

Dig out the top spit of soil from the second strip and turn it onto the top of the first trench. Add peat to this at the rate of one 10 litre bucket per metre (one 2gal bucket per yd) and further quantities of steamed bonemeal, fishmeal and powdered seaweed at the same rate as before. This method cultivates the soil to a depth of 46cm (18in) with the most fertile soil remaining on top and the best possible base preparation underneath.

If your soil is predominantly sand or gravel, it is advisable to add extra humus forming material and, if possible, to line the bottom of the trenches with old grass turves, turned upside-down. If your soil is shallow, above a limestone subsoil, the only practicable way to grow roses is to build up the soil above ground level with abundant quantities of peat and farmyard manure. This prevents constant contact between the high pH soil and the rose roots.

Allow the soil to settle for at least one month after preparation and before the arrival of the new bushes. If pathways run between the beds, remove the top few inches of soil from the pathways and place on top of the rose beds. Raised beds tend to warm up and drain more quickly.

While waiting for your new rose bushes, mark out the beds for total area and position, using canes for the bushes and string or garden twine for the rows.

With a bed width of 1.4m (4ft 6in), grow the roses in three rows 46cm (18in) apart, with 23cm (9in) gaps between the outside rows and the edges of the bed. The distance between each bush should be 46cm (18in) for Large Flowered roses and 61cm (24in) for Cluster Flowered and Polyantha roses. Plant Miniature roses 30cm (12in) apart with 30cm (12in) between rows, or grow them in 15–18cm (6–7in) pots, plant tubs or troughs.

Before planting Shrub roses, consult a specialist book to find out the height and vigour of your chosen variety. Then devise your bed layout according to the space needed by each plant.

First-class bed preparation requires considerable effort and determination but provides a sound basis upon which to build. 'Wonder' fertilisers are often cited by exhibitors when they produce top-class blooms, but no amount of feeding will produce winning blooms consistently, without a healthy root system.

Planting

Plant during October or November while the soil is still friable and retains some warmth (though it is possible to plant up to early April).

When the bushes arrive, they are usually packaged in strong brown-paper sacks, which are waxed inside to retain moisture. Unpack the bushes, leave them tied in bundles and immerse the roots in water for twenty-four hours to give them a good drink.

If the bushes are almost completely dehydrated, with wrinkled roots and stems, dig a deep hole in the garden and bury the bushes completely. Soak the soil and leave the bushes for ten days. When the bushes are lifted the roots and stems should have returned to their original healthy state and they can be planted.

Before planting, examine the bushes for broken stems and roots, which should be cut back cleanly. If any strong thong roots are longer than 30cm (12in), shorten them, and remove any leaves. When planting bushes, keep types and varieties together to simplify cultivation. Dig the holes to suit the root systems of each individual bush, as they all differ in shape.

Line each hole with 2–3 handfuls of a planting mixture, made from 9 litres (2gal bucket) of damp peat mixed with a large handful each of bonemeal and seaweed powder. Hold the bush in the hole so that the budding union (where the stems of the cultivated variety emerge from the wild rose rootstock) is at soil level or not more than 2.5cm (1in) below. Fill in the hole with soil, adding several handfuls of the planting mixture.

Firm the soil around each bush, using the ball of the foot and working from the outside up to the stem. The surface of the bed should be level when you have finished the planting operation.

Winter Treatment

Fallen leaves should be gathered up and burnt or destroyed as they can carry overwintering spores of diseases such as mildew, blackspot and rust.

During November, shorten the longer stems on all types of bush roses by about one third, to combat wind-rock which will loosen the bushes and cause gaps that may fill with water and then freeze. Loosened bushes should be refirmed after filling the gaps around the roots with fine soil. Towards the end of December, if all the leaves have not fallen, pull them from the bushes and burn them.

After this, spray all roses, paths and beds with a winter wash, using a solution of Jeyes Fluid in water – 5tbsp per 4.5 litres (1gal). This will kill off most of the overwintering spores of mildew, blackspot and some dormant insects and their eggs, giving you a cleaner start to the season.

Weather permitting, each year apply a surface dressing of powdered seaweed to all rose beds at the rate of 68–136g per sq metre (2–4oz per sq yd). This will improve fertility and soil structure. Farmyard manure should also be obtained, stacked, and allowed to rot down for use as a mulch later.

4 Pruning to Spraying

Before you start pruning, have ready two pairs of sharp secateurs, short and long handled. There are numerous makes of secateurs, but probably the best are made by Felco, Wilkinson and Rolcut. Felco secateurs are preferred by most professional nurserymen. Keep the cutting blades sharp, well oiled and correctly aligned. Poor cuts lead to bruised stems and can result in stems dying back, or disease entering damaged areas.

Pruning times depend on geographical location and the related winter conditions, plus the need to time the blooming of varieties to correspond with particular show dates. Bushes that are exposed to searing northerly or north-easterly winds are often subject to severe wind burning. If stems are frost or wind damaged almost to soil level there will be little healthy wood left to prune.

Also, varieties have different lengths of time between pruning and blooming – this time may vary from about thirteen to eighteen weeks. Seek advice from local growers about the usual blooming and pruning dates of your varieties in your area. However, remember that this advice can only yield estimated dates and, therefore, it is advisable to stage-prune banker varieties. For example, if twelve bushes of a particular variety are to be pruned, cut back six on the estimated pruning date and the remaining six ten days later. This will give staggered blooming. Furthermore, if stems are pruned lightly 15–20cm (6–8in) above soil level, they will produce blooms more quickly than if cut to 2.5–5cm (1–2in). This can be used to help stagger bloom production, on bushes with a sufficient length of healthy stem.

As bushes are planted closely together, cut out any weak, thin, straggly or dead wood to avoid congestion, leaving just four to six healthy stems. Cut into clean, healthy, white wood at the predetermined height above ground, about 5–6mm ($\frac{3}{16}$–$\frac{1}{4}$in) above outward-pointing buds. The cuts should be sloping downward at 45° away from the bud.

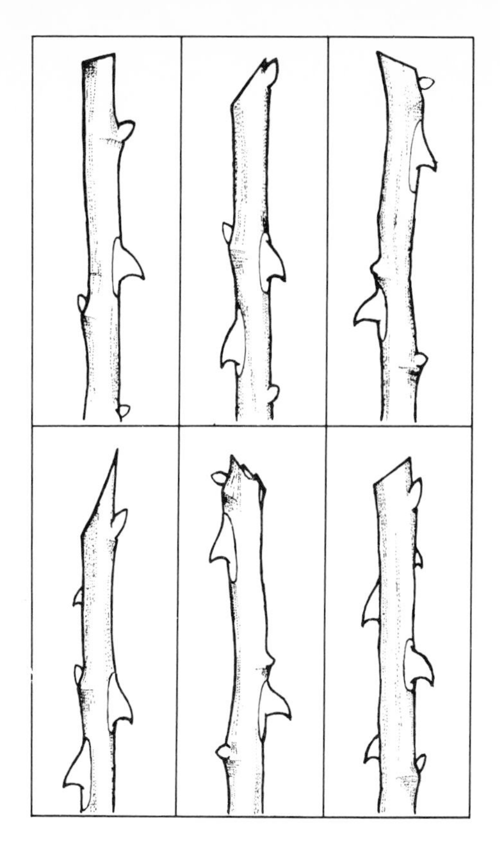

Pruning cuts: (a) too high; (b) too close to bud; (c) sloping wrong way; (d) cut too long; (e) cut jagged; (f) correct cut

If bushes are badly damaged by frost, you have little choice but to cut back the stems until you reach good white pith, free from any brown marking which would indicate frost damage. I have had to cut back many bushes to ground level in the past, but despite this very few have died (there are more dormant buds nearer to the budding union).

Prune newly planted bushes first, starting during the last two weeks in February but avoiding periods of hard frost. Cut all stems back to 5–8cm (2–3in) above ground level.

Prune established Cluster Flowered varieties next, starting during the last week in February. Prune the previous year's growth, which developed from strong base shoots, to 20–30cm (8–12in) above soil level, but cut back older stems to 5–8cm (2–3in) above ground level. Again, cuts should be made above outward-pointing buds.

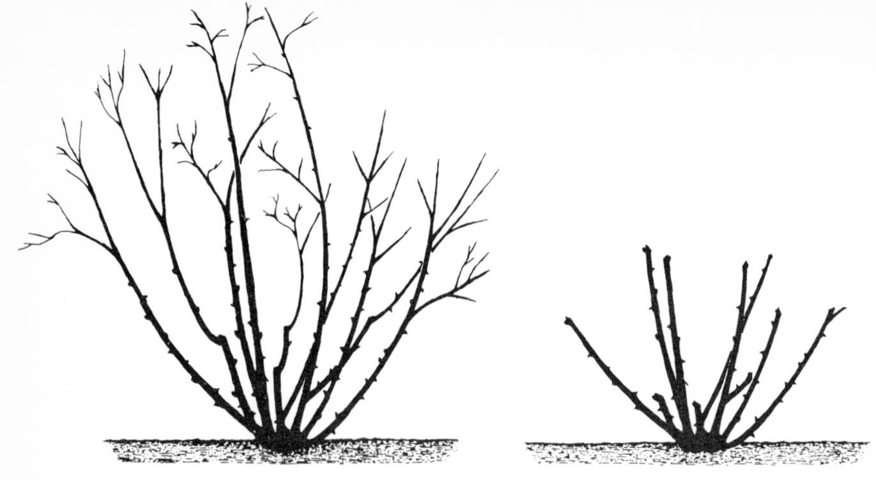

Established Cluster Flowered variety before and after pruning. New stems are lightly cut back, old stems are well cut back

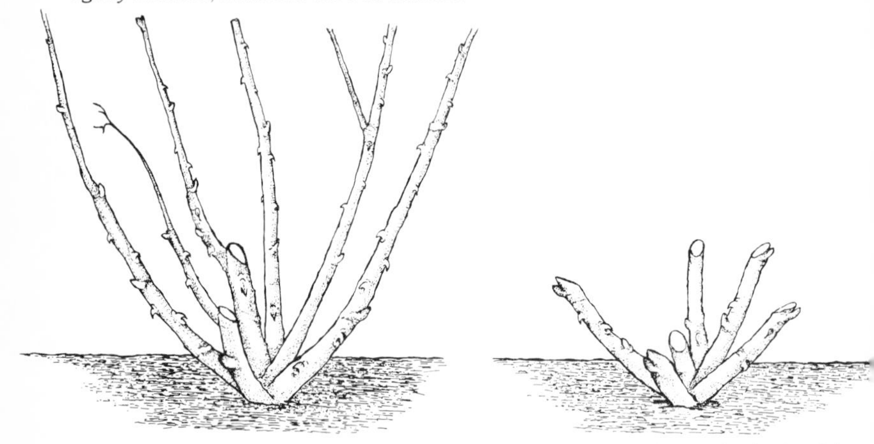

Large Flowered variety before and after pruning. Note that the top buds should always point outwards

Large Flowered varieties should be pruned at predetermined dates. Prune first those varieties with the longest pruning-to-flowering period. Stage-prune bushes if necessary.

The pruning-to-blooming period for Miniature roses may be only ten to twelve weeks, so they should be pruned last, cutting all healthy wood back to 5–8cm (2–3in) above soil level.

Shoot and Bud Restriction

Five to six weeks after pruning, the remaining stems will produce a number of shoots from previously dormant buds. These should be restricted to five or six selected shoots per bush, evenly spaced, preferably pointing outward. Rub out unwanted shoots with your thumb or forefinger. Many modern varieties also produce axil shoots at either side of the main shoots. These should also be rubbed out as soon as possible. Without restriction there will be great congestion and competition – and you will not produce top-size blooms.

From early June, Large Flowered varieties form flower buds at the tips of the growing stems. Some varieties produce side buds just below the terminal bud, or numerous side shoots carrying buds from leaf axils. Remove these gradually over several days, to allow the bush to channel its energy into producing larger blooms.

Cluster Flowered varieties produce stems with clusters of buds at the top. With these stems, remove the terminal bud as soon as possible, allowing surrounding buds more space to develop. The object is to produce a circular mass of evenly spaced blooms. Often, you will need to remove other excess buds to allow adequate space for flower development. Occasionally, if blooms are too early for a particular show, the first cluster of buds should be allowed to develop, then the resultant flowers and their footstalks should be removed cleanly. The remaining flower buds on the stems will then open, providing possible show trusses.

Any stems not providing flower buds are called blind, and should be cut back to half their length to encourage new growth from a dormant bud.

Suckers from the wild rootstock can appear at any time during the growing season. Although they often emerge from the soil some distance from the rootstock, they are easily identified as their foliage and stems differ considerably from the cultivated varieties. Suckers grow rapidly, absorbing plant food and energy, and should be removed at their point of origin as soon as is practicable. Use a hand fork to remove soil along the sucker's course and wrench or cut it from its socket.

The Main Rose Flushes

Large Flowered, Cluster Flowered and Miniature roses provide blooms in two main flushes. The first flush occurs in early summer, the second in late summer and early autumn (usually September). The period between the flushes varies with the prevailing weather conditions and the geographical location, but it is usually between

(*left*) Shoot restriction. Remove excess side shoots with thumb and forefinger
(*right*) Removing axil buds to produce one large bloom per stem

ten and fourteen weeks. The main flushes of bloom last for three to four weeks and provide the majority of blooms for exhibiting (most shows are planned to coincide with them). However, there is always an intermittent supply of blooms between the main flushes, from new basal shoots which emerge and develop very quickly during this period.

After the first flush, cut back the remaining rose stems as soon as they have flowered. They should be cut back to half their length, at a point just above a leaf joint where there is an outward-pointing bud. This will encourage new stems to develop ready for the autumn shows. If stems are shortened less, subsequent growth will be inadequate in length for exhibition purposes.

Occasionally, when the first flush of blooms occurs too late for the summer shows, flowering stems should be cut back before blooms are fully developed, to encourage the second flush. Between the main flushes, follow your normal cultivation routine. In addition, water and foliar feed as necessary if conditions are very dry or wet.

During this period, most rose bushes develop new basal shoots, which carry blooms for the current season and provide new wood for the following season. These new shoots are easily dislodged by high winds, so they must be secured to canes inserted nearby. This is essential with the large heads of bloom on Cluster Flowered varieties.

Between flushes, cut out thin straggly growths which have developed around the base of bushes, and also remove and burn any discoloured, damaged or diseased leaves. This encourages air circulation and discourages insects and disease.

Feeding

Roses respond positively to good initial soil cultivation and subsequent controlled feeding. After pruning, a dressing of 135g per sq metre (4oz per sq yd) of fishmeal should be applied to the beds and hoed into the top 25mm (1in) of soil. Two weeks later, apply a further dressing of fertiliser, this time a good proprietary rose fertiliser (eg PBI Toprose, ICI Rose Plus, Fisons Rose Fertiliser); or, my own preference, a 4:1 mixture of two slow-release fertilisers, Enmag and Gold N. The latter is a nitrogen fertiliser to encourage new stem and leaf growth. This second application should be at 100–135g per sq metre (3–4oz per sq yd). Distribute the fertiliser evenly around the bushes and again hoe it into the top 25mm (1in) of soil.

Feed twice more during the season: at the beginning of July and of August, with 35g per sq metre (1oz per sq yd) applications of a proprietary rose fertiliser. This will not be necessary if you used a slow-release fertiliser initially.

Many rose exhibitors also apply foliar feeds at regular intervals (seven to fourteen days), and may combine such feeds with insecticides and fungicides to save spraying time. This is acceptable, providing that only compatible constituents are used and that they are mixed thoroughly; otherwise foliar scorching may occur. Advice on spray compatibility is often given on bottle labels. Also, ICI now market Roseclear, a combined insecticide and fungicide which can be used with foliar feeds.

A foliar feed applied on its own during very wet or prolonged dry periods certainly helps the roses along, providing nourishment when the roots are unable to absorb it, or when it is simply not available at root level. I find that two good foliar feeds are Maxicrop (liquid seaweed extract) and Fillip, a balanced liquid inorganic chemical feed.

Liquid feed may be applied to roots to enhance bloom size and colour. A liquid rose or tomato feed is suitable for this purpose, but it should be applied only two or three times during the period prior to each flush, when the buds are swelling. This particular type of feeding is very hard work when over 250 bushes are being grown.

Over many years, my feeding programme has consisted mainly of Enmag and fishmeal – and my results have always proved highly satisfactory.

Mulching and Watering

Following the application of fertiliser, all rose beds should be given a mulch of old, rotted farmyard manure, peat, spent hops or good garden compost. This provides a cool, moist blanket over the fine feeding roots, preventing them from drying out during periods of drought. The organic material will gradually decompose further, be carried down into the soil by the worm population and provide additional humus.

Peat and hops do not contain any plant nourishment, though farmyard manure and garden compost do provide some sustenance for the roses. However, constant mulching of the beds improves the texture of the soil and thus its ability to retain and use fertilisers when they are applied. Mulching is a very important stage in providing a top-class environment for long-term bloom production.

Mulching will help to conserve water, but during really dry periods it may well be necessary to water the bushes at the roots. Use a hose pipe or garden sprinkler, or, for smaller numbers of bushes, a watering can. The beds should be thoroughly soaked. Allow 4.5–9 litres (1–2gal) of water per bush at each watering.

During periods of drought when water is restricted it is particularly beneficial to spray bushes once or twice per week with a weak foliar feed. This will supply through the leaves nutrients which are unobtainable from the soil. Watering and foliar feeding during hot dry summers tend to produce the largest and most colourful blooms.

Pests and Diseases

All varieties of roses are susceptible to pests and diseases to some extent, so, to produce prizewinning blooms, you must provide bushes with adequate protection. The main insect pests are greenfly, thrips and earwigs, although damage is also done by leaf rolling sawfly, cutter bees, chafers, capsid bugs and caterpillars.

Common rose diseases are mildew, recognisable by a white powdery appearance on leaves and stems; blackspot, identified by the small, fringed black spots on the lower leaves; and rust, evident as orange spores, usually on the underside of the leaves in early summer, which turn black in the autumn. Rust was regarded as a killer disease until recently, but it can now be eradicated by Plantvax, a new fungicide.

If in doubt about a particular pest or disease, refer to one of the many publications which provide a more thorough means of identification, covering all the pests, diseases and symptoms in greater detail than is possible here.

As always, prevention is better than cure, and the best way to prevent damage is to spray at regular intervals (ten to fourteen days), beginning in mid June and concluding in late autumn. Use a preventative spray in fine mist form, from a suitable sprayer. Many types of sprayer are available, so use one with a suitable capacity. Growers with up to 250 bushes will require a 4.5–9 litre (1–2gal) capacity unit. Above this number, a knapsack sprayer with a capacity of 13–18 litres (3–4gal) would be more suitable.

Never spray in bright sunshine, as droplets left on the leaves magnify the sun's rays and result in burnt foliage. Preferably, spray in the evening (in still conditions) so that the bushes can assimilate the spray overnight.

Over recent years, many systemic insecticides and fungicides have appeared on the market, but I have tried many over the past twenty years and most have proved almost totally ineffective.

My present spraying programme uses Nimrod T systemic fungicide (combats mildew and blackspot) together with Sybol II insecticide (kills greenfly and small caterpillars). After several combined sprays of Nimrod and Sybol II, or Roseclear, it would be wise to use different types to prevent an immunity building up. Murphy's Tumblebug and Tumbleblite provide my current alternatives to the main spraying programme.

This programme gives protection against greenfly, the main sap-sucking insect; caterpillars, which eat leaves and flowers; and blackspot and mildew, which can cause defoliation or severe damage to leaves and stems.

Earwigs and large caterpillars should be discouraged from climbing stems and eating into the blooms by smearing petroleum jelly around the stem 15cm (6in) below the bud.

Thrips can also be a menace at flowering time. These tiny black insects are usually prominent in hot dry weather, attacking white, yellow and pastel-coloured roses in preference to the darker varieties. They cause brown marks on the flowers and eat petal edges, leaving them ragged. The main remedy is to spray with malathion, but *never* mix this with any other ingredient as it will lead to leaf-scorching.

Sawflies lay eggs on the leaves and then inject poison into them, causing the leaves to roll up around the eggs and form a natural protection until the eggs hatch out. Quick removal and destruction of the affected leaves is probably the best remedy.

Foliage can also be destroyed by the rose slugworm sawfly larvae which eat the soft tissue between the leaf veins, leaving only a 'skeleton'. This slugworm is yellow in colour and can be destroyed by a spray containing fenitrothion. Both of the sawfly adults are small shiny black flies.

The froghopper (or cuckoo-spit insect) and the leaf hopper, which jump away quickly if you attempt to touch them, are further pests to the rose grower. They attack leaves and soft new growing tips, sucking out the sap and causing malformation of stems and leaves. To eradicate the problem, spray affected bushes with products containing menazon, dimethoate or pirimiphos-methyl.

Providing that you maintain the spray programme, most of these minor rose enemies can be handled easily, as they occur.

(*above right*) City of Gloucester. Large Flowered specimen bloom of classic shape. This example displays good colour and would grace any box, bowl or vase exhibit

(*below right*) Europeana. Cluster Flowered rose displaying good truss form and colour

EXIT
EXIT
BARB

5 Bloom Protection

Large Flowered (Hybrid Tea) roses grown for exhibition generally require protection from rain or strong sunshine so that they may be presented to the judges in a near-flawless, fresh condition. Rain marks the delicate petals of most roses, and may even cause the petals to stick together and rot; the blooms then often completely fail to open. Strong sunshine can bleach out the colours of many varieties, so the blooms need to be shaded, although some varieties, eg Grandpa Dickson, City of Bath and Silver Jubilee are almost immune from weather damage and bleaching.

Cluster Flowered (Floribunda), Polyantha (Pompon), Miniature and Shrub roses do not require protection.

Protectors should be placed in position over opening blooms, just when the buds show colour and the sepals begin to part. A wise exhibitor will keep a wary eye on the weather forecast, and cut blooms early to avoid any bad weather which may be predicted for the day before a show. If the weather is wet during the week before a show, all but Large Flowered varieties may be cut two to three days before a show. Shake the rain from the heads and stand the stems in water indoors, where many varieties will continue to open.

The vast majority of rose exhibiting in the British Isles is conducted under rules laid down by the Royal National Rose Society. The rules governing amateur exhibitors growing roses outdoors require that the bushes must have been owned and grown solely by the exhibitor for three months prior to a show. The roses must have been grown in the open, and only individual bloom protectors may be used. This therefore outlaws any form of blanket protection. Note, however, that not all shows use RNRS rules and you should enquire for any additional or alternative regulations.

(*above left*) Twelve stems of Miniature rose Starina, daintily arranged and well presented in a small urn. Awarded first prize at the Lakeland Show in 1978
(*below left*) Display of new roses staged by the British Association of Rose Breeders at the summer centenary show of the RNRS in 1976

Commercial covers were produced until recently and there are still thousands being used. If you get a chance to obtain them from retiring exhibitors, take it.

Home-made Covers

Many exhibitors manufacture their own bloom covers because of the high cost (or unavailability) of the commercial products. Conical covers can be made easily with materials which are generally available. You will need strong galvanised or plastic-covered wire to secure the cone to the supporting stake or cane. Plastic-covered wire is preferable, as this will remain unaffected by the weather. Most hardware stores, garden centres or agricultural merchants can supply suitable coils of wire.

The actual covers can be made from a variety of materials such as thick cartridge paper or plastic sheeting. The latter can be from empty fertiliser bags, or bought from agricultural merchants, garden centres or specialist sheet-plastic suppliers. Cartridge paper should be painted with a waterproof paint before assembly.

You will also need rolls of 25mm (1in) wide PVC adhesive tape or masking tape, a small stapling machine, a medium-sized pair of pliers, and stout canes or stakes 12–16mm ($\frac{1}{2}$–$\frac{5}{8}$in) thick and 1.2–1.5m (4–5ft) long.

Mark out circles 50cm (20in) in diameter on the material chosen for the cone manufacture. Cut them out, then into semi-circles. Form semi-circles of material into a cone by drawing both ends together, allowing a 10–12mm ($\frac{3}{8}$–$\frac{1}{2}$in) overlap. Fasten two staples through the overlap to hold the cone together. Cover the entire length of the joint (on the inside) with a single piece of adhesive tape.

Cut a 1.1m (3ft 6in) length of wire from the coil. Then, using pliers, form a 19–20cm (7$\frac{1}{2}$–8in) diameter circle in the centre of the wire, with two wire 'tails' each 25–28cm (10–11in) long. Place the wire circle over the outside of the cone and allow it to slide down until it will rest unsupported, with the wire tails on the cone joint. Mark the position where the wire tails touch the joint, remove the wire and make a 6mm ($\frac{1}{4}$in) diameter hole at this point. Push the two wire tails through the hole from the inside until the wire ring rests firmly against the inside of the cone. Use four or five strips of adhesive tape, each about 10cm (4in) long, to secure the ring to the inside of the cone.

Next, at a position 5cm (2in) from the cone, bend the twin wire

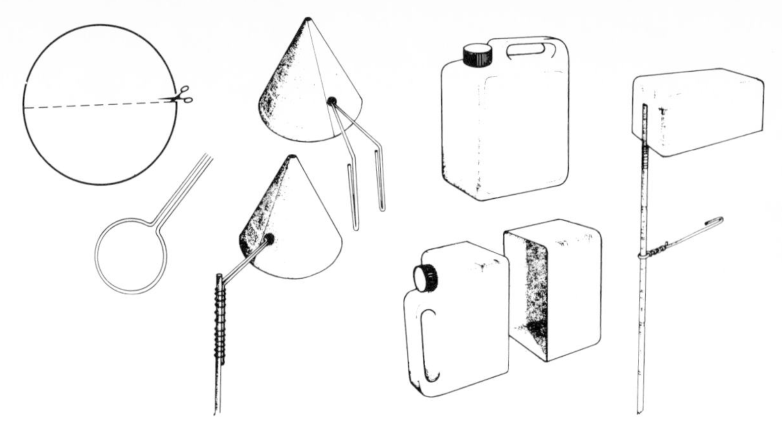

Making a conical cover Making an emergency cover

tails vertically downward, then back upon themselves. Finally, bind this wire 'handle' to the top of a cane or stake.

Although this procedure sounds fairly complex, with a little practice you will be able to produce these very serviceable covers fairly quickly.

Emergency bloom covers can be made from the large 4.5 litre (1gal) plastic containers used for detergents or bleaches. They should, of course, be thoroughly washed out first.

When all other supplies of covers have been exhausted, you can use 13–15cm (5–6in) plastic plant pots, with their drain holes covered.

Plastic Bags as Protectors

During recent years some exhibitors have experimented with small 200 x 250mm (8 x 10in) plastic bags as bloom protectors. I have found these useful, but only for certain varieties. Only the very light, opaque and flimsy bags should be used, as heavier bags tend to bend the bloom necks and also rub and bruise the petals. Bags provide a much cheaper protection than covers; they save time; and (when sealed) they create a damp, warm, local environment in which blooms tend to develop more quickly than they would under covers.

However, bags also have their disadvantages, the chief one being that the damp internal atmosphere can cause petal rot or loss of colour in some varieties. Another major disadvantage is that you cannot observe the development of the bloom. Bearing these advantages and disadvantages in mind, I have used bags experimentally for five years, and have reached the following conclusions.

Bags produce excellent results with some varieties when used from the bud to three-quarters developed bloom. Good varieties are Gavotte, City of Gloucester and Admiral Rodney.

Some varieties seem totally unsuitable for bag covering. These are usually the varieties with only small numbers of large petals, eg Red Lion, Grandpa Dickson and Embassy.

Many slow-opening varieties which have many petals can be covered for about seven days in a bag and then finished under covers. This method produces the best results with varieties such as Red Devil, My Joy and Bonsoir.

If the weather is damp and foggy, blooms may rot even under covers. In such conditions bags provide a means of producing some good blooms when other means would fail.

Bags can be useful when blooms are opening too slowly.

Using Covers

Once placed over a bud, the cover's height should be adjusted so that the opening bud cannot be damaged if the stem is blown about in a wind, though make sure the cover still gives protection.

When commercial covers are used, altering the height is extremely easy. Less sophisticated, home-made covers are adjusted simply by pulling the cane slightly out of the bed until the protector is in the required position. As blooms open, the covers may require further adjustment.

When the cover has been correctly positioned, there is a further safety precaution to be taken. Twist one end of some thin plastic-covered wire securely around the cane and fasten the other end, gently but securely, around the stem of the bloom, about 20–25cm (8–10in) below the base of the bloom. Such precautions should eliminate bloom damage in all but the most severe of wind conditions.

Should a bloom develop a split or snub-nosed centre (see page 7), remove the cover and use it elsewhere, as these blooms will be useless for show purposes.

If a bloom is large and apparently suitable for specimen classes, you can tie the centre to help hold its shape. Run two turns of thick, soft wool around the centre cone of petals and then tie it loosely with a double twist, locating twists away from the petal edges to avoid cutting them.

(*left*) Conical cover in position over a bud just showing colour
(*right*) Plastic-bag protection for a promising bud

Using Plastic Bags

Make sure that the neck of the opening bud has stiffened before it is bagged. Gather the open end of the bag together and inflate it by blowing into the gathered opening. Slip the bag over the bud and secure it at the gather, about 40mm ($1\frac{1}{2}$in) below the base of the bud, using a paper/wire tie.

Buds take from nine to fourteen days to develop from just showing colour to three-quarters developed. A wise exhibitor should check the development by opening the bag occasionally. When a bag is opened you will see that a considerable amount of moisture has been retained, so turn the bag inside out before replacing it.

Whichever form of protection you choose, make sure that the greatest possible number of blooms are protected before a show. In our changeable climate, there is no doubt that covering blooms greatly enhances the chances of winning.

6 Cutting and Transporting Blooms

Roses take up most of their moisture during the hours of darkness and lose it during the heat of the day. Thus, the best times to cut roses are early on the day of the show, when the stems, leaves and flowers are fully charged with sap taken up during the night; or the previous evening, so that the stems can be stood in cold water overnight.

However, if circumstances dictate it, roses can be cut at any time of the day, as long as they are immediately stood in cold water. Cluster Flowered and Shrub roses should be cut first, Large Flowered varieties next and Miniature roses last. This system reflects the different cut lifespans of the various types of roses.

Before cutting, make sure that you have enough buckets. These should be of sufficient depth to allow cut rose stems to be immersed in 20–25cm (8–10in) of water, and should have wide bases so they are not easily overbalanced. You can buy special flower buckets with side handles, but suitably sized, ordinary plastic buckets will usually serve the purpose just as well.

Cut roses should be stood in a cool, dark place. In such a situation, blooms will develop slowly and remain fresh for up to two days. Garages, large sheds or cellars are suitable standing areas. Keep the temperature as low as possible and exclude light as far as you can. There should be sufficient space for the roses to be stood without crushing and to provide easy access for bloom sorting.

Cutting Method

Before starting to cut, place a flower bucket containing 15cm (6in) of cold water nearby. Using a sharp pair of secateurs, cut about 45–60cm (18–24in) below the bloom, just above a leaf joint and preferably where the axil bud is pointing outward.

Place the cut stem immediately into the waiting bucket. Continue cutting the other roses, but avoid overcrowding the stems in the bucket, as this can cause petal damage or bruising. Once the bucket is full, take the stems to the standing area and transfer them to the other previously filled buckets. To avoid damaging blooms and foliage, separate the stems carefully at this stage.

As cut blooms no longer have a root system through which to absorb water, beware of exposing them to sunshine or wind, which will hasten the dehydration of blooms and foliage.

If roses have to be transported some distance from the growing plot to the standing area, you need some means of preventing the buckets from tipping over whilst in transit. One idea is to construct a framework of wooden hurdles into which the flower buckets can be secured. It should have three or four compartments, each wide enough to hold a bucket securely. Another idea is to use a lightweight plastic milk or beer crate and plastic bottles (eg squash or washing-up liquid bottles) with their tops cut off. Once filled with water and placed in the crate, the bottles provide a safe way of transporting the roses.

At the standing area, sort the cut roses into separate varieties. Where possible, each variety should be stood in a separate bucket to assist sorting at a later stage.

The roses should then be allowed to take up water for at least three to four hours before the final sorting and transportation to the show. During this time, if any of the blooms show signs of wilting (indicating that they have not taken up water), recut the stem, about 25mm (1in) from the bottom, while still under water.

Sorting Blooms

When the roses have had a good drink, they should be re-examined critically for any faults which may have been overlooked when cutting. Looking first at Large Flowered blooms, those with split or snub-nosed centres (see page 7) should be rejected completely. Old blooms or blooms which show signs of bad weather marking or petal damage should also be discarded.

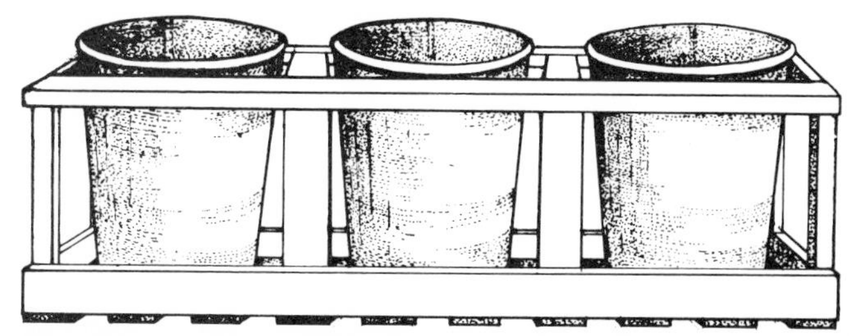

Framework for carrying buckets to the standing area, after cutting blooms

You should be left with fresh, well-formed blooms of varying sizes. At this point, after reference to the show schedule, any specimen blooms should be selected from the remaining roses and put to one side.

Secondly, Cluster Flowered roses should be examined, one stem at a time. Any dead, badly marked or damaged individual flowers should be neatly removed, together with their footstalks. Any excess buds which are not showing colour should be similarly removed from the truss.

Shrub roses should be taken to a show as they are cut from the bush, with only damaged blooms being removed.

Finally, Miniature roses should be checked, and damaged blooms should be discarded or removed from the truss.

Transporting Blooms to the Show

When travelling to a show, blooms may have to be transported over distances ranging from one to several hundred miles. No matter what distance is involved, you must have a safe, practical system if roses are to arrive at the show venue in the same condition in which they left home. Give the subject some attention well before the show season, as it requires considerable thought and preparation, depending on the number of roses to be carried and the distance to be travelled.

The rose stems should be carried in water at all times, and blooms should not be allowed to rub against each other in transit. It should be easy to insert and remove stems from the carrying container, to prevent bloom and foliage damage by thorns raking against petals and leaves.

When only a small number of roses are to be exhibited locally, blooms can be transported in plastic flower buckets, in 5–8cm (2–3in) of water. Attach a 10cm (4in) wire grid to the top of the bucket to separate stems. Prevent rubbing during transit by placing soft tissue paper between each bloom. When a dozen or more stems are to be carried over a distance greater than two or three miles, an alternative carrying method should be used.

The crate and bottle method is cheap and easy to organise and use; it is useful if you are showing between forty and eighty blooms locally. Each bottle will take up to two stems. To prevent stems from moving in transit, insert newspaper wedges into the bottles after the stems have been put in.

Purpose-built carrying crates provide the best means of transporting large quantities of blooms

An exhibitor wishing to compete at regional or national shows will need to transport many blooms over long distances. In this case, to utilise your vehicle's capacity to the full, you will need purpose-built carrying crates, which have proved very successful. The roses stand in tubes slotted into holes drilled in the crates.

Before constructing a crate, measure the space available inside your vehicle, bearing in mind that most modern saloon and estate cars have sides which slope in slightly towards the roof, and sometimes at the rear. The overall height of a carrying crate with blooms is about 61cm (24in), and you should also allow about 8cm (3in) around each side of a crate. Allow space for exhibition boxes, flower buckets, and the various other smaller items of staging equipment (see page 47).

It is usually possible to fit two carrying crates into a car (larger estates can carry up to four). This gives a carrying capacity of between 100 and 200 stems, which should prove adequate for even the most demanding of show schedules.

Packing blooms into crates

Cut the sorted roses to length, bearing in mind the vertical space available inside your vehicle. It is best to carry stems as long as possible – they can always be shortened later if necessary. To assist the blooms to take up water during the journey, the bottom 8cm (3in) of each stem should be stripped of bark, leaves and thorns, using a sharp knife. This also makes it easier to insert the stems into the tubes.

When packing roses into crates, keep the varieties together as much as possible, to simplify the task of sorting and comparing blooms of a given variety at the show. As each individual stem is inserted into a tube it should be tied, using a paper/wire tie, to the adjacent pea-stick rail. This operation requires great patience and care, but will ensure that blooms cannot rub against one another in transit.

Transporting Cluster Flowered, Polyantha and Miniature Roses

Cluster Flowered and Polyantha rose stems may be transported in crates as described above, but you should take additional measures to prevent bloom damage in transit. Having secured the stems to the rails of the crate, a loop of strong wool should be tied firmly around each truss to prevent the blooms hanging too far over the crate. Providing the wool is tied carefully, this will do no damage.

As these types of roses have large trusses they tend to require a greater amount of water, so remove them from the crates and place them in buckets of water as soon as possible when you arrive at the show.

The much smaller size of Miniature roses means that they cannot be carried to a show using any of the methods mentioned so far. An ideal way to carry them is to insert the stems into a damp block of Oasis, which is inside a plastic container (eg an icecream or margarine tub) to retain the moisture. These containers will not take up much room in the vehicle, and will be safe and steady.

7 Show Preparation

Before an exhibit can be staged, a variety of equipment has to be manufactured or bought and certain preparations must be made. Most of this should be done well before the start of the show season, though some has to be left until you actually get to the show. Remember to obtain your show schedules early, scrutinise them in detail and take them to the relevant shows.

A show schedule usually requires roses to be displayed in vases, bowls and boxes. Many shows provide vases and bowls, which should be used, but some shows expect exhibitors to provide their own show-bench equipment. Before the show season starts, consult the schedules and then assess your requirements accordingly.

Vases and bowls are easy to obtain through advertisements in the popular gardening press or from specialist manufacturers. The bikini-type vase, with detachable base, is the most popular with exhibitors. It is also the most practical, as vases can be stacked one inside another when not in use. The sizes most often required by show schedules are 20cm (8in) and 25cm (10in). Turning to bowls, those with 18cm (7in) and 25cm (10in) diameters are most often required.

Very few shows provide exhibition boxes and, unfortunately, they are not readily available. Exhibitors competing in box classes usually manufacture their own boxes.

Rules governing box sizes are laid down by the Royal National Rose Society and these are generally accepted by most show committees. Boxes for specimen blooms should all be 102mm (4in) high and 457mm (18in) wide. They should be 305mm (1ft) long for 6 blooms, 610mm (2ft) for 12 blooms and 1,067mm (3ft 6in) for 24 blooms. There should be 127mm (5in) between the centres of each tube in the same and adjacent rows. It is not difficult to make a box but it does require a little time and patience, and is, therefore, best done before the start of the show season. To construct a six-bloom box, you will require:

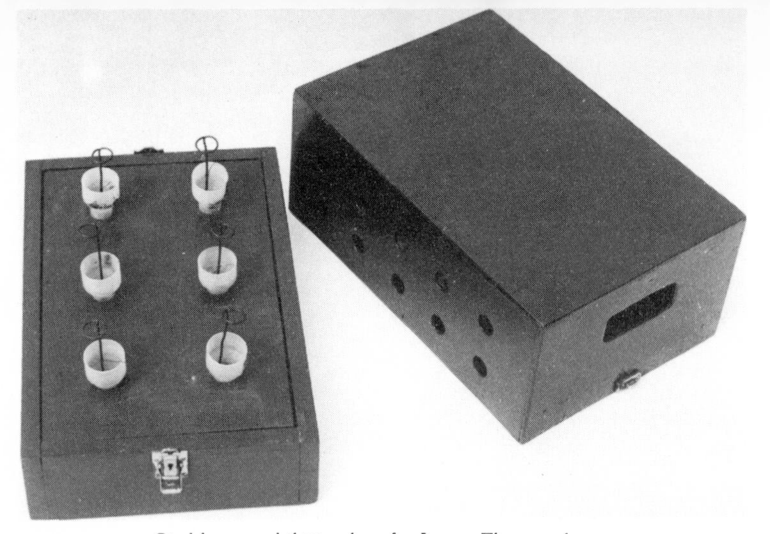

Six-bloom exhibition box for Large Flowered roses

1.52m (5ft) length of 102 x 13mm (4 x ½in) softwood (planed and dressed)
610 x 457mm (24 x 18in) of 5mm (¼in) plywood or hardboard
3m (10ft) length of (½in) squares softwood
nails (25mm and 13mm/1in and ½in)
wood glue

Cut the 102 x 13mm (4 x ½in) softwood into two 457mm (18in) and two 28cm (11in) lengths. Glue and nail them into a rectangle with outside measurements of 457 x 305mm (18 x 12in). Next, place the frame on top of the plywood or hardboard, and mark out two rectangles the same size as the inside of the frame. These rectangles should then be cut out and placed to one side.

The 13mm (½in) square softwood should then be cut into lengths to make supporting battens which fit inside the frame, near the top and bottom. The battens should be glued and nailed in position so that the two pieces of plywood will rest on them and fit flush with the frame, forming a closed box. Fix the bottom piece of plywood into position using glue and 13mm (½in) nails.

The centre position of each tube should then be marked out on the top piece of plywood, with 127mm (5in) between each centre in each direction. Tubes can be stomach or denture tablet tubes, but they must be cut down to only 127mm (5in) in length. Alternatively, you can make tubes by cutting 25mm (1in) diameter plastic tubing into 127mm (5in) lengths and glueing plastic discs to the base of the tubes. The plastic tubing can be obtained from hardware stores or plumbers' merchants.

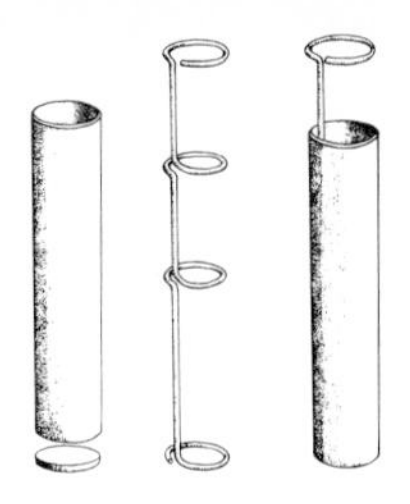

Tubes with bloom supports for exhibition boxes

Whatever tube is used, drill holes in the top piece of plywood, of a size that will easily take the tubes, but not allow them to move about. The top fits loosely against the supporting rail, so that it is easily removed.

Fit each tube with a bloom support, made from 12 or 16swg copper or plastic-coated wire. Bend a 36cm (14in) length of wire using pliers, to form three hoops with a straight piece joining them. One hoop should be at the bottom of the support, one at a height of 75mm (3in) and the final hoop at a height of 15cm (6in). The hoops should be wide enough to fit snugly into the inside of the tubes. Finally, paint the boxes dark green using an oil-based paint.

If more than one box has to be transported to a show, construct a strong, lightweight lid from softwood and plywood so that they can be stacked. Cut large ventilation holes and carrying handles out of the lids, and fit catches with integrated locks if possible (useful if boxes are returned by road or rail after two or three day shows). Larger exhibition boxes can be constructed using exactly the same method.

If Miniature roses have to be shown in a box, it should be 17.2cm (6¾in) long, 11.4cm (4½in) wide, 8.9cm (3½in) high at the back and 5.1cm (2in) high at the front. These can be obtained from the RNRS at St Albans (see Introduction).

Other Equipment

The rose bowls provided by many shows come complete with a grid which is supposed to hold stems in place, but it tends to turn the staging operation into more of an engineering feat than an art. Discard these grids and use a staging medium instead. Blooms can be staged in cut reeds or soaked newspaper, but the best staging medium is Oasis.

Take two flower buckets to each show, to fill your vases with water. They are also useful for checking and comparing blooms. You will also need a sharp long-bladed knife, to remove bark and thorns and cut Oasis; a sharp pair of secateurs; a small pair of scissors, for removing or trimming damaged petals and leaves; a packet of 30cm (12in) rose wires to straighten bloom necks; a small pair of pliers, to cut the wires; a camel-haired brush and a supply of cotton-wool balls, to dress blooms; and a small jug or watering can to top up vases after staging. In addition, take a reel of fine flower-arranger's wire, a black waterproof marking pen, a supply of variety name cards and a retractable measuring tape to each show.

On Arrival at the Show

If staging is to take place during the hours of darkness, it is advisable to find a position on the bench directly under an artificial light source, if possible (and not too far away from a water supply). Once unloaded and installed in your position (boxes should be on the floor, safely out of the way), fill both buckets with water and soak the Oasis – it can take up to fifteen minutes for a block to become saturated.

Obtain any vases and bowls from the equipment distribution point, and use the measuring tape to check vase sizes. This is important because, if the wrong vase or bowl is used, you will be disqualified. One or two spare 25cm (10in) vases are useful later when blooms have to be compared or dressed. Then, obtain your class and exhibitor's entry cards from the show secretary and check them against the schedule for accuracy. Should you find any discrepancies, consult the show secretary.

Next, all the blooms in the carrying crates should be untied and allowed to stand freely in their tubes. Do not rush this operation as it requires a great deal of care. Any bloom or truss which appears to be drooping should be removed from the carrying crate and placed in a bucket. Cut off the bottom 25mm (1in) of stem while it is under water, then leave it in the bucket to take up water.

By this time, the Oasis should be saturated. Remove the blocks from the buckets, and allow any excess water to run back into the buckets. Place the blocks on the staging bench and press the tops of the vases and bowls firmly onto the blocks to leave an impression of their shape. Cut the Oasis to shape using the long-bladed knife, then place it into the vases and bowls.

You are now ready to start staging.

8 Staging

Before you can impress the judges you must understand what they are looking for in your blooms. Judges generally use the RNRS standards, under which points are awarded for the various constituent parts of each exhibit. Whatever the type of rose, the emphasis is always on the quality of a bloom or truss, assessed by form, size, substance, freshness and colour. Additionally, points will be awarded for healthy, green foliage and well-balanced stems (except for specimen blooms in boxes). Further points will be awarded for presentation.

Up to one third of the available points can be awarded for foliage, stems and presentation, so prizes can be won or lost by good or bad staging. To become a top-class rose exhibitor, you must have a skilful staging technique.

Dressing Blooms

The object of dressing a bloom is to open a young bloom earlier than nature intended, to remove any slight imperfection from an otherwise near-perfect bloom, or to manipulate the petals into a slightly better position. Dressing should only be practised occasionally, where absolutely necessary, as overdressed blooms will be down-pointed by the judge.

Generally, it is only practical to dress Large Flowered blooms, as other types of rose tend to have fewer petals and any dressing would be apparent immediately. Before starting any type of dressing, remember that petals are extremely delicate and very susceptible to bruising.

Young blooms

Half-open blooms may not display their full form and beauty, so inserts may be placed between individual petals to open the flower. Use multi-coloured cotton-wool balls for inserts; colours which contrast with the petals will be seen more easily when they have to be removed later.

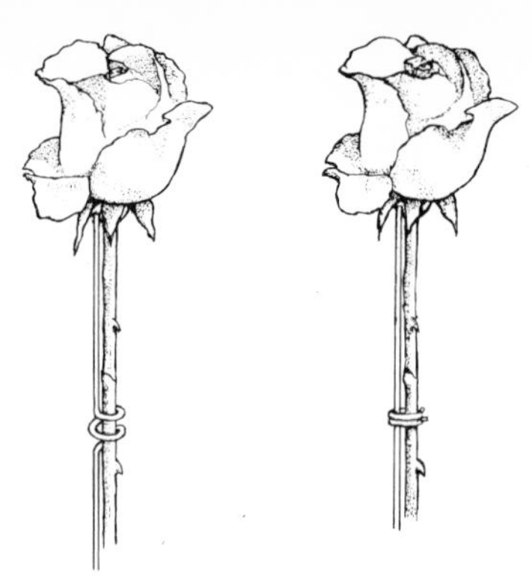

Two wiring methods. The wire must enter the seed cup to support the bloom

Push the outer petals gently outward and downward, and insert the cotton-wool balls to hold the petals in position. The inner petals should be manipulated slightly outward to fill any gaps. Also, the main outer ring of petals should be encouraged to reflex at the edges: use the first finger of one hand to support the petal from below, and the thumb and forefinger of the other hand to mould the edges downward. Finally, loosen the central petals by blowing gently into the bloom. Thus, you will have produced a larger bloom, displaying its full beauty and form.

This size increasing technique can also be used to create a specimen bloom, although you should beware of opening a bloom beyond the optimum point. If the central petals are loosened or opened too much, the bloom loses its regular, pointed centre.

This type of dressing may be done before leaving for a show or when staging exhibits at the show. Remember to remove any cotton-wool before judging, otherwise the exhibit will be down-pointed.

Folded petals

Blooms often have a fold on the centre petals. Use a camel-haired brush to tease out the fold, by applying pressure to the inside of the fold and gently brushing outward.

Damaged or misshapen petals

If a bloom has one or two damaged petals, pluck the offending

(*right*) Winning vase of six Large Flowered specimen blooms staged by Mr T. G. Foster at the centenary autumn show of the RNRS in 1976. Included is the silver-medal-winning rose Fred Gibson (top right)

CLASS 19

ST. ALBANS
THE ROYAL NATIONAL ROSE SOCIETY
NATIONAL SHOW
FIRST PRIZE
CLASS
12
NAMES OF VARIETIES
Royal Highness

Red Devil
Anne Letts

(*left*) Young bloom dressed with cotton-wool to open the flower
(*right*) Three-stage bloom exhibit

petals neatly out of the flower and manipulate the surrounding petals gently to fill the gap. If the damage is only on the outside edge of the petal, trim it neatly with a small pair of scissors. The petal should then be shaped so that the edge turns under, away from the judge's eye.

Wiring

Occasionally, Large Flowered rose varieties produce good flowers on stems which are slightly bent or too weak to support the weight of the bloom. In such cases, the bloom heads may be supported by discreet wiring.

Insert a stiff 30cm (12in) florist's wire into the seed cup at the base of the bloom. Great care should be exercised as bloom heads are snapped off easily. Push the wire gently up towards the bloom,

(*above left*) Bowl of twelve Royal Highness stems, displaying fresh, unmarked, classically shaped blooms with clean, healthy foliage. An excellent example of the art of arranging bowls

(*below left*) Box of twelve distinct specimen blooms, with the Lakeland Bowl. The blooms are at the peak of their development (*Mrs M. Kirk*)

thereby straightening the stem. Once the stem has been straightened, the wire and stem should be held firmly together, just below the footstalk. Twist the wire firmly around the stem at this point. Twist the remaining wire around the stem at a lower point. Remove any excess wire with a pair of pliers.

Alternatively, after the wire has been pushed up into the seed cup, straighten the stem, hold it next to the supporting wire, then wind a piece of fine flower arranger's wire around both the stem and the support wire, just below the footstalk. Repeat further down the stem. This method reduces the risk of snapping off the bloom head at the footstalk.

Box Exhibits

Schedules may require six, twelve or twenty-four specimen blooms of Large Flowered roses to be exhibited in boxes. The schedule will also state the minimum number of varieties which must be used in the exhibit. The boxes should comply with the RNRS specifications and should be constructed as described on page 46.

As always, points are available for presentation. The top of the box around the tubes should be covered before staging, preferably with fresh damp moss. Not only does this give a natural green background to the exhibit, but it also creates a damp atmosphere, which helps to keep blooms fresh.

Also before staging, fill the bloom tubes in the boxes with cold water, and fasten 7.5cm (3in) lengths of fine flower arranger's wire to the bloom supports, about halfway between the top of the tube and the top of the bloom support. Now turn to the blooms.

Examine the largest specimen blooms for any major flaws which may have developed (eg split centres or snub-noses). If necessary, blooms should be dressed at this stage. Recheck the minimum

A bloom tied with soft wool to prevent the petals opening

number of varieties required by the schedule, then make the final selection of blooms.

In general, the roses should be arranged in a box in size order, starting with the largest bloom in the top left-hand tube and working down until the smallest bloom is in the bottom right-hand tube. However, in multi-variety classes, you should try to blend bloom colours. Blooms of similar colours should not be placed next to each other; and, where possible, the whites, creams and soft pastel shades should be used to cool the bright vermilions, yellows and deep pinks. So, in the interests of good colour blending, the rule about sizing may be ignored, on occasions.

Once the bloom positions have been finalised, the stems should be cut down to approximately 20cm (8in) and the leaves removed. The stems should then be cut to their final size, allowing the blooms to rest with the calyx inside the bloom support. Turn the blooms to present their best face forward, and secure them in position using the wire attached to the bloom support.

Box exhibits tend to be rather time consuming to stage, so, if you have a long way to travel, or a large number of exhibits to stage, it may be best to stage the boxes before leaving home. If boxes are staged at home, the centre of each bloom should be tied loosely with two turns of thick soft wool, to avoid the petals loosening during the journey. On arrival at the show, remove the wool immediately and replace any damaged or fully opened blooms.

Vase Exhibits

Vase exhibits, containing up to nine blooms or trusses, are the major part of most show schedules. The schedule will state the vase size to use, and vases should be checked before exhibits are staged.

Vase exhibits are always viewed from the front and should be staged with this in mind. They should be well balanced, with both blooms and foliage displayed to their best advantage. Large Flowered blooms should be spaced so that they do not touch each other, but without large gaps between the flowers. Cluster Flowered and Polyantha roses should be staged so that the trusses fit together to form a circular or oval outline, without any of the individual blooms being crushed together too tightly.

The height and width of an exhibit should balance with the size of the vase and the blooms or trusses. If a 25cm (10in) vase is used, the exhibit should be approximately 63cm (25in) high, and if a

20cm (8in) vase is used, the exhibit should be approximately 51cm (20in) high. The width of an exhibit should be linked to bloom or truss size, eg six specimen Large Flowered blooms, in arcs of three blooms, should occupy a width of 30–38cm (12–15in). (Blooms are usually 10–13cm/4–5in across.)

Consult the schedule to find out the number of varieties required for each exhibit. If specimen blooms are required, use your largest flowers and remove any side buds. If specimen blooms are not stipulated, use average-sized, bright, colourful blooms, each carrying up to two side buds.

Bloom patterns

You choose your own arrangement of blooms or trusses but most exhibitors follow set patterns. The following patterns are those which I have used successfully for Large Flowered roses, but they can be adapted for Cluster Flowered or Polyantha roses.

Three blooms in a vase should be arranged to form an equilateral triangle, with two blooms at the top and one at the bottom.

Four blooms should be arranged in a square or diamond shape.

If you have six blooms, insert two at the same height at the rear of the vase, then make an arc of three blooms below these, the central bloom in the arc being equidistant from the two rear blooms. Insert the sixth bloom at the front of the vase, directly below the central bloom of the arc.

If you have nine blooms, insert two at the rear, then arcs of three and four blooms under these.

In all cases, foliage should be neatly arranged, right side up. Any damaged or marked foliage should be neatly removed, and the vases should be wiped clean.

Multi-vase classes

For these classes the schedule will stipulate the number of vases, the number of blooms per vase, and whether a separate variety is to be staged in each vase. These exhibits are usually displayed on tiered staging, and the aim should be to produce a continuous slope of colour. The top bloom or blooms in each vase should be slightly below the lowest bloom or blooms on the next tier. If any variety to be exhibited possesses particularly short stems, adjust the height of the other vases accordingly.

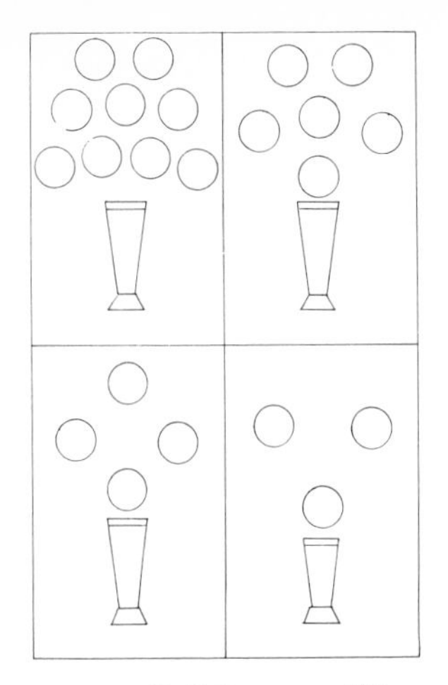
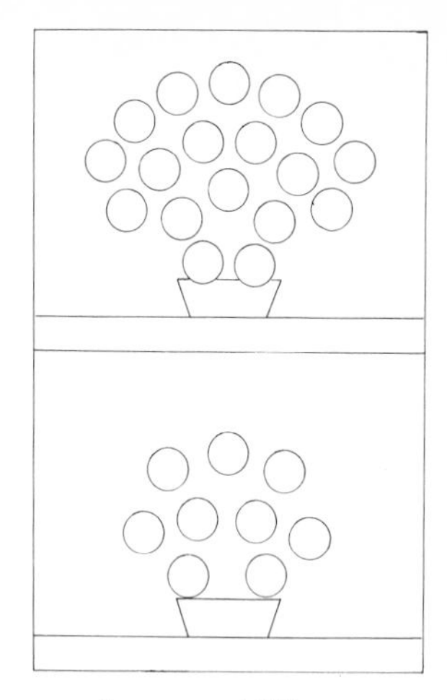

(*left*) Suggested bloom patterns for vase exhibits
(*right*) Suggested bloom patterns for bowl exhibits

Three-stage bloom classes
This type of exhibit has three blooms of the same variety at varying stages of development. There should be one bud, one perfect bloom and one fully open bloom (see page 53). The bud should be showing colour with one or two petals beginning to unfurl above open sepals. The perfect bloom should be circular in outline, fresh and of good colour. The fully open bloom should be fresh and circular in outline. Stamens, if visible, should also be fresh.

Arrange the blooms in a vertical line, with the tallest being the bud, at the rear of the vase. Insert the perfect bloom below this, then the fully opened bloom at the bottom, at the front of the vase, with the lower petals obscuring the rim of the vase. Foliage should again be neatly arranged.

Bowl Exhibits

The arrangement of roses in bowls is always regarded as a major challenge by exhibitors. This is probably because they provide the finest and most eyecatching method of display. All types of roses can be displayed in bowls. Schedules will stipulate the number of stems and varieties required. They will also state whether an exhibit is to be arranged for 'all-round' or 'frontal' effect.

The 'all-round' bowl of roses should be almost spherical, and equally attractive from all sides. For these exhibits, about 25mm

(1in) of Oasis should be left protruding above the rim of the bowl.

The 'frontal' bowl of roses should be round or oval when viewed from the front. This is by far the most common type.

Large Flowered roses ('all-round' bowl)

Fresh, average-sized blooms with up to two side buds per stem should be used for bowl exhibits.

Start with the tallest bloom, at the centre of the bowl. Add more blooms in circles around this central bloom, working outwards and downwards, in ever widening rings until you reach the diameter required. Then continue inwards and down, in reducing circles. When you reach bowl level, insert the stems horizontally so that they hide the bowl rim. Foliage should be neatly arranged, with any surplus or damaged leaves removed.

Cluster Flowered roses ('all-round' bowl)

The arrangement is similar to the one used for Large Flowered roses. The aim is to achieve a spherical mass of colour, with no large gaps between trusses. Faded, damaged or marked blooms, and any buds not showing colour, should be neatly removed, together with their footstalks.

Large Flowered roses ('frontal' bowl)

The schedule will stipulate the number of stems required, which is most commonly nine, twelve and eighteen.

Use average-sized blooms and remember to blend colours when more than one variety is required. Blooms should not touch each other, but there should be no large gaps between flowers. When staging this type of exhibit, most exhibitors follow one of many set patterns. Most of these have arcs of blooms, with the highest arc at the rear of the bowl and the lowest arc at the front.

My successful pattern arranges a typical bowl of twelve stems in three arcs of blooms. There are four blooms at the rear, five in the centre and three at the front. The final three blooms are inserted at almost bowl-rim height. Foliage should be neatly arranged, with any surplus or damaged leaves removed.

Cluster Flowered roses ('frontal' bowl)

Again, the number of stems required will be stated in the schedule. Before starting to arrange the stems, remove any faded, damaged or marked blooms and any buds not showing colour, together with their footstalks. The pattern should be similar to that for Large Flowered roses in 'frontal' bowls. The aim is to produce a colour-balanced, circular or oval mass of blooms with no large gaps between each truss. The whole arrangement should be slightly convex when viewed from the side.

Miniature Rose Exhibits

For the sake of simplicity, I have so far ignored the exhibition of Miniature roses, but they too are exhibited in boxes, vases and bowls.

Bloom size varies greatly, approximately 7–40mm (¼–1½in) in diameter, according to variety. Avoid showing the extremes of bloom size together, as this will give your exhibit an unbalanced appearance. Judges do not award points for bloom size, so, blooms of any size can be exhibited in the same class, against each other.

Box exhibits
These exhibits should be staged in Miniature rose boxes, the dimensions of which are given on page 47. Points are awarded to each bloom for form, freshness and colour, and each box of six is awarded points for presentation.

Select evenly sized blooms and cut them down so that they rest on the rims of the box's tubes. Think about colour blending as you proceed with the staging.

Vase exhibits
Schedules may require one, three, five or six stems to be displayed in vases, and 7–10cm (3–4in) miniature vases will usually be provided by the show. Each stem should bear a single truss of blooms.

The stems should be arranged, in Oasis, in the same way as Large Flowered blooms in vases. Foliage should be neatly arranged.

Bowl exhibits
Schedules will require a given number of stems of a given number of varieties to be displayed in miniature bowls, urns or round blocks of Oasis placed upon a saucer (receptacles will usually be provided by the show). Bowls or urns should be filled with saturated Oasis, protruding 2cm (¾in) above the rim.

Before staging, examine each truss and remove any marked, damaged or badly faded blooms, together with their footstalks. Arrange the trusses like Cluster Flowered blooms in a 'frontal' bowl.

Shrub Rose Exhibits

The arrangement and display of this type of rose is less formal than for the other types. Emphasis is placed upon a large quantity of blooms in natural sprays or clusters.

On the Bench

Once your staging is finished, consult the schedule again and check the number of stems and varieties in each exhibit. Then check each bloom and remove any cotton-wool dressing or wool ties. Remove any insects, using a camel-haired brush. Before placing the exhibits on the show bench, top them up with water, and wipe all receptacles clean and dry.

The exhibits may now be placed on the show bench, taking care to place them in the right areas for their classes. Exhibits should be positioned squarely, facing forward (except 'all-round' bowls) and should not touch adjacent exhibits. If possible, position them so that the judge can look directly into the blooms, and so that they are not obscured by other exhibits. Box exhibits should be tilted at the back by 5cm (a couple of inches) – a block of Oasis is the ideal support.

Finally, place the variety name card (face up), and the exhibitor's card for the relevant class (face down), in front of the exhibit; and remove all your equipment.

9 Judging

Newcomers to exhibiting should realise that many judges have themselves been novice exhibitors, and are very willing to discuss your problems or doubts. Many times, after judging, I have rearranged exhibits in the novice section, dressed blooms or offered advice that might help for the future.

During the last ten years, the RNRS has established a countrywide nucleus of competent, qualified judges. High standards were set and have been maintained. Thus, there now exists a list (from the RNRS Secretary) of specialist rose judges, who are usually happy to judge upon request.

If judging, you must find out the extent of your duties, whether you are judging alone, and the availability of stewards. Obtain schedules early and check details: RNRS rules usually apply, but sometimes Royal Horticultural Society rules are followed. Take along the appropriate rule books, and a copy of the RNRS *Rose Directory*, listing over 700 varieties by type, name and colour. Arrive early at the show venue, contact the show secretary, avoid involvement with exhibitors and start the judging punctually. When judging alone, with possibly only a steward for assistance, first check the general standard of the exhibits by walking quickly around the rose benches. Next, locate an example of a good average-sized, Large Flowered bloom of any variety, to be used for reference purposes.

Judging usually starts with the specimen bloom classes, because these blooms must be at their peak, and they often pass this stage quickly in warm exhibition tents and halls.

The judge checks the number of exhibits in each class, and that they are on the correct staging. (This should be verified by a steward.) Next, the eligibility of exhibits is checked, including that the correct sizes of boxes, vases and bowls are being used (if these are stipulated), that the number of stems and varieties is right, and that the varieties themselves are eligible ones. Infringement of these rules leads to disqualification and the dreaded NAS (Not As Schedule) being marked on your exhibition card. One tip to judges is to check the availability of bowls and vases at the show. Shows often run short of the correct sizes and allow substitutes, but forget to inform the judge.

After checking eligibility, the judge dismisses all obviously inferior exhibits. Sometimes this eliminates all but two or three exhibits, and, even then, an experienced judge can place the remaining exhibits in order of merit without pointing. The judge only has a difficult task when there are many exhibits of a similar standard.

Generally, the rose classes are divided as follows:

Large Flowered (HT) Roses
Specimen Blooms, exhibited in boxes, vases and bowls.
Large Flowered Blooms, exhibited in vases and bowls.
Three Stage Blooms, exhibited in vases.

Cluster Flowered (Floribunda type) and Polyantha (Pompon type) Roses
Exhibited in vases and bowls.

Miniature Roses
Exhibited in boxes, vases and bowls.

Judges are occasionally requested to judge Trade and Floral Art exhibits. This should not be undertaken unless the judge has past experience, or is being assisted by another experienced judge.

Large Flowered, Specimen Bloom Classes

The judge is specifically looking for large, fresh, well-shaped blooms of a good colour, which is representative of the variety. A good bloom can be awarded 3 points, but exceptional blooms can earn an additional 1 or 2 points.

If a bloom is left tied or pelleted, 3 points are deducted. Serious bloom faults, such as poor shape (split or snub-nosed centres), weather damage or tiredness (which shows as dull colour or limp petals) incur penalties of up to 3 points.

Foliage is not required or considered for box exhibits. Two extra points may be awarded per six-bloom unit for good presentation, eg uniformity of size, good colour blending with a pleasing background, and neat labelling of varieties. A box of six specimen blooms could thus be awarded 18 points, plus 2 points for presentation, totalling 20 points. This total would be increased if any exceptional blooms were present.

In vase and bowl classes, extra points are awarded under two headings: stems and foliage, and presentation (3 points for each). Stems should be straight and well balanced in relation to the blooms they carry (ie not too thin or thick); foliage should be ample and healthy. In these classes, presentation points are awarded to well-balanced, graceful arrangements, taking into account their overall height and width. Blooms should predominate, being suitably spaced and colour blended. Foliage should be neatly arranged.

Large Flowered Bloom, Cluster Flowered and Polyantha Classes

Exhibits in these classes should give an overall impression of bloom, youthfulness, freshness and bright colour. Each receptacle in these classes is considered an individual unit and may be awarded a total of 30 points: two bowls can earn 60 points, and three vases can earn 90 points. Points are allocated as follows:

Form, size and substance of individual bloom(s) or cluster(s)	10
Freshness, brilliance and purity of colour	10
Stems and foliage	4
Presentation	6
Total	30

Large Flowered individual blooms should be of good, average size and shape. Clusters should be round in outline and as large as possible. Stem, foliage and presentation standards are the same as for specimen blooms in vases and bowls, but the proportion of points is significantly higher in these classes.

Three points are deducted for each Large Flowered bloom left tied or pelleted. Undersized, overdressed, misshapen, tired, dull, marked or damaged blooms will be penalised.

Points will be deducted from Cluster Flowered roses for trusses of poor outline or with large gaps between blooms; overcrowded individual flowers with many unopened green buds; dead or old blooms, footstalks or hips left in the truss.

Large Flowered, Three Stage Bloom Exhibits

Points are awarded on exactly the same basis as for other Large Flowered classes.

If a bloom is left tied or pelleted, 10 points are deducted. Overdressing of blooms, poor arrangement and poor foliage also lead to points being deducted.

Miniature Rose Classes

Miniature roses are displayed in boxes, vases and bowls, and classes are divided accordingly. The object is to produce a dainty, artistic display.

For box exhibits, use fresh, bright, average-sized, well-shaped blooms, of circular outline, with regular well-shaped centres. If blooms display stamens, these should be fresh. Blooms should be neatly positioned, and colour blended, where necessary. Each bloom can be awarded 3 points, with an additional 6 points available for presentation. Penalties are the same as for larger roses, but I have never deducted points for tied or pelleted blooms.

Points for vases and bowls are the same as for Large Flowered or Cluster Flowered classes. Schedules call for a given number of stems, which may carry single blooms, or clusters in various stages of development. They should be arranged to produce dainty, balanced, gracefully artistic displays. Laterals which have developed subsequent to main stem growth (a predominant habit) are not allowed.

You will benefit greatly from studying the show benches after judging. Examine both winning and losing exhibits and try to assess where points were won and lost. If possible, take photographs of the winning exhibits, as they will teach you more about staging than many words of text.

10 The Rose Exhibitor's Calendar

This is a month-by-month reminder of the most important tasks. It is based on conditions in the north-east of England, so dates will be about four weeks earlier for south-west England and two weeks later for north-east Scotland. It is also useful to have long-term records, so keep a notebook covering feeding, spraying, pruning and assessment of varieties.

January

1 Continue planting new bushes and rose rootstocks, when soil is not too wet or frosted.
2 Give all rose bushes and beds a winter wash, using Jeyes Fluid.
3 Join a local rose society and the Royal National Rose Society if you have not already done so, or renew membership.
4 Obtain supplies of fungicides, insecticides, fertilisers and Oasis in readiness for the coming season.
5 Make bloom protectors.

February

1 Continue planting rose rootstocks, when ground conditions permit.
2 Refirm soil around any bushes loosened by frost.
4 Start pruning newly planted bushes and Cluster Flowered varieties during last two weeks of the month. Delay this operation during periods of hard frost.
5 Head back rootstocks budded during previous season.

March

1 This is the last month for planting new bushes safely; prune stems before doing so, to save stooping later.
2 Continue pruning Cluster Flowered roses. This should be completed by the end of the second week.
3 Start pruning Large Flowered varieties, following previously estimated dates. Try to complete pruning by end of the month.
4 Firm soil around bushes loosened during pruning.
5 Dispose of prunings by burning if possible. Do not compost, as prunings may carry diseases upon them.

April

1 Prune all Miniature roses during the first week.
2 Apply fertiliser dressings to rose beds.
3 Mulch beds with well-rotted farmyard manure, peat, hops or garden compost, preferably after rain, to trap in the moisture.
4 Spray remaining rose stems with fungicide to kill off any overwintering blackspot, mildew or rust spores.

May

1 Thin out any excess shoots emerging from base of bushes.
2 Check new growth for signs of greenfly or small caterpillars.
3 If required, start spraying with a suitable insecticide.
4 Weed beds regularly.
5 Obtain show schedules as early as possible.
6 Make a final check of show equipment.

June

1 Start regular spray programme against pests and diseases (every 2–3 weeks).
2 Water, if necessary, when conditions are dry.
3 Liquid feed for extra bloom size and colour.
4 Check for, and cut back, blind shoots.
5 Start disbudding of flowering stems. Remove side buds of Large Flowered varieties and truss centre buds of Cluster Flowered varieties.
6 Position bloom protectors or bags over buds when they show colour.
7 Obtain a supply of fresh moss from a woodland or riverside and keep this fresh by watering occasionally.

July

1 Apply dressing of rose fertiliser, and water in if conditions are dry.
2 Protective spray must be applied during the first few days of the month. This will avoid the need to spray again until the first main flush of roses is over.
3 Continue bloom protection; check bloom development daily.
4 Tie promising roses which may produce specimen blooms.
5 Cut blooms required for exhibition on the evening before the show or the morning of the show.
6 Cut back stems not used for show, immediately after flowering.
7 Bud rootstocks after rain, or water thoroughly two days before budding.

August

1 Apply final dressing of rose fertiliser during first two weeks of the month.

2 Continue spraying programme; change constituents to avoid insect and disease resistance building up.
3 Foliar feed if conditions are very dry or very wet.
4 Water if necessary.
5 Check on new varieties at shows. Obtain new catalogues and order new roses as soon as possible for the coming season.
6 Start preparation of new rose beds towards end of the month.

September

1 Complete preparation of rose beds.
2 Last month for budding rose rootstocks.
3 Protect blooms ready for autumn shows.
4 Send off rose orders (nurserymen may soon run out of stock).
5 Obtain a supply of peat ready for planting new roses.

October

1 Continue protective spraying until the middle of the month.
2 Plant new bushes, which should start arriving at the end of the month.
3 Start cutting back long stems to prevent wind rock during winter.

November

1 Most new roses arrive this month. Plant as soon as possible.
2 Dispose of winter prunings and fallen leaves (preferably on Guy Fawkes' bonfire).
3 Obtain supplies of manure, and stack under cover, if possible, to prevent nutrients being washed out by winter rain.

December

1 Plant new bushes when soil is not frozen or too wet.
2 Plant rose rootstocks, if required for coming season.
3 Service pruning equipment.
4 Check, repair or service show equipment. Order or make new items required.

Index

Page numbers in italic indicate illustrations